Business Intelligence and Human Resource Management

Business Intelligence (BI) is a solution to modern business problems. This book will discuss the relationship between BI and Human Resource Management (HRM). It will also discuss how BI can be used as a strategic decision-making tool for the sustainable growth of an organization or business. BI helps organizations generate interactive reports with clear and reliable data for making numerous business decisions. This book covers topics spanning the important areas of BI in the context of HRM. It will also discuss the aspects, tools, and techniques of BI and how it can assist HRM in creating a successful future for organizations. Some of the tools and techniques discussed in the book are analysis, data preparation, BI-testing, implementation, and optimization on GR and management disciplines. It will include a chapter on text mining as well as a section of case studies for practical use. This book will be useful for business professionals, including but not limited to, HR professionals and budding business students.

Business Intelligence and Human Resource Management

Concept, Cases, and Practical Applications

Edited by

Deepmala Singh

Symbiosis Centre for Management Studies (SCMS), Nagpur
Symbiosis International (Deemed University)(SIU), Maharashtra

Anurag Singh

School of Business, University of Petroleum and Energy Studies (UPES), Dehradun

Amizan Omar

University of Bradford, UK

S.B. Goyal

Director, Faculty of IT, City University, Malaysia

A PRODUCTIVITY PRESS BOOK

First published 2023
by Routledge
605 Third Avenue, New York, NY 10158

and by Routledge
4 Park Square, Milton Park, Abingdon, Oxon, OX14 4RN

Routledge is an imprint of the Taylor & Francis Group, an informa business

ISBN: 9781032027302 (hbk)
ISBN: 9781032027296 (pbk)
ISBN: 9781003184928 (ebk)

DOI: 10.4324/9781003184928

Typeset in Garamond
by codeMantra

Contents

Foreword

The capacity to use information to achieve a competitive advantage over competitors is referred to as business intelligence (BI). Businesses all over the world consider BI as the most desired technology. Even in these difficult economic times, when IT expenditures are being slashed, BI remains a top priority for executives. BI brings together people skills, technologies, applications, and business processes to help make better strategic and tactical decisions. Data management methods for planning, collecting, storing, and structuring data into data warehouses and data marts, as well as analytical tasks such as querying, reporting, visualizing, generating online active reports, and running advanced queries for clustering, classification, segmentation, and prediction, are among the activities and applications. Data marts are limited to a specific process or department, such as the Human Resources (HR) department, while data warehouses focus on enterprise-wide data.

At this juncture, this research book *Business Intelligence and Human Resource Management: Concept, Cases, and Practical Applications* is timely and will play a vital role in examining the managerial aspects of BI systems in various domains. It creates a set of rules for generating value via the use of BI systems and technology. The book focuses on BI as a process for serving complex information demands in creating insights and providing decision support driven by a combination of human and technological skills. This book covers everything from the definition of BI, its value creation, its meaning and quality, its depiction and management of both internal and external data, its presentation and visualization, and analytical strategies and future predictions. The author provides practical views on the increasingly popular "big data" theme for the future. This book unearths all the key variables in great detail that must be examined to successfully harness the value of intelligence for both novices and professionals.

This book isn't just for BI professionals but also for the data community in general. The explanations of data requirements identification, data modelling and design, data quality, integrity, mining, warehouses, and other important issues are clearly presented and valuable. In general, it is readable, comprehensive, and exhaustive for practitioners and presents a wealth of ideas that can be used by enterprises.

The contributions of the book's editors, Dr. Deepmala Singh, Dr. Anurag Singh, Dr. Amizan Omar, and Dr. S.B. Goyal are commendable.

My best wishes for your book! Happy reading!

Dr. Mani Venkatesh
Head of MSc in Big Data & AI Program
Associate Professor
Department of Strategy and Entrepreneurship
Montpellier Business School

Preface

Business intelligence (BI) can be thought of as a prediction tool that makes use of both current and historical data. It is a methodology, a set of processes that analyze raw data in order to produce meaningful information that may be used to develop an effective plan for gaining a competitive edge over competitors. BI enables businesses to make informed decisions. It consists of the following components: analysis, data preparation, data architecture, BI testing, implementation, and optimization.

BI might be viewed as a panacea for contemporary business challenges. Global business trends are always evolving as a result of cultural change and technical improvement. Organizations are likewise altering themselves to keep up with the pace in order to obtain a competitive edge over their competitors. To adapt to any change, comprehensive analysis and data-driven procedures are required. BI enables organizations to develop interactive reports that contain accurate and reliable data that can be used to make multiple business choices. This book compiles a variety of approaches to BI in human resources and other management domains. It will assist readers in developing a knowledge of the movement of business decisions from old approaches to more analytical and applied management strategies. Additionally, it will provide insight into the use of business information as a tool for strategic decision-making in order to ensure the organization's long-term viability. It covers a wide range of human resource and other management-related issues. This book discusses BI approaches and technologies such as analysis, data preparation, data architecture, BI-testing, implementation, and optimization in the human resources and management disciplines.

This book is organized into well-structured chapters written by academic and industry researchers from around the world who are all experts in their respective fields of study. Each chapter focuses on a single topic, namely how to leverage BI tools and methodologies in HR and other management areas to create a projected future for the firm. Today's managers benefit from having a competitive edge over their competition. Its application is visible in the day-to-day planning of manpower, the differentiation of product offerings, the knowledge of consumer behaviour, and the development of policies to ensure the organization's viability.

Deepmala Singh, Anurag Singh, Amizan Omar, S.B. Goyal

Acknowledgements

The editors of *Business Intelligence and Human Resource Management: Concept, Cases, and Practical Applications* appreciate the writers' efforts in submitting their excellent chapters to our edited text book on time. We would also like to express our gratitude to Michael Sinocchi, Publisher – Business Improvement, for his unwavering support and direction throughout the book's production.

I would like to acknowledge with gratitude the love and support of my family: my father, Krishna Murari Singh; my husband, Santosh Singh; my daughter, Ritisha; and my sisters, Manisha and Anamika. I wish to thank all reviewers for providing constructive feedback to the writers in order to improve the chapters' quality, coherence, and content presentation. This publication would not have been possible without the input of reviewers. Also, I would like to thank my co-editors, Dr. Anurag Singh, Dr. Amizan Omar, and Dr. S.B. Goyal, for their unwavering support and belief in me. I thank everyone who assisted us directly or indirectly in the completion of this edited book.

Deepmala Singh
Lead Editor

Editors

Dr. Deepmala Singh is an Assistant Professor at the Symbiosis Centre for Management Studies (SCMS) in Nagpur, Maharashtra. She worked as an Assistant Professor at the LBEF Campus in Kathmandu (in academic collaboration with Asia Pacific University Malaysia). In 2016, she earned her PhD from Banaras Hindu University. Her study concentrated on BHEL's human resource development practices. Prior to dedicating her time to LBEF, she worked as an Assistant Professor and Guest Faculty at prestigious management colleges such as MNNIT Allahabad. In addition, in 2011, she worked as a project fellow on a significant research project supported by the UGC. Her research interests include Strategic Human Resource Management and Organizational Behavior. She has published more than 20 articles in prestigious national and international (A category and SCI) journals.

Dr. Anurag Singh is currently working as a Senior Assistant Professor in the area of Human Resource Management and Organizational Behavior with School of Business, University of Petroleum and Energy Studies, Dehradun. Prior to this he worked with Institute of Advanced Research, Gandhinagar, and Baba Saheb Bhim Rao Ambedkar Central University, Lucknow. He has also worked on a short-term research project in Dubai on CSR with Arabia CSR Network, Dubai. He completed his doctorate from Banaras Hindu University. He is a post-graduate in MBA (HR), M.Com., M.A. (Economics). He has qualified the prestigious Indian UGC-Net in two different subjects, i.e., Human Resource Management and Commerce. He has also cleared Company Secretaries, Executive level. He has published more than 10 research papers in various national and international journals and presented his research work in more than 17 national and international research conferences.

Dr. Amizan Omar is the Interim Director of Accreditation for the Faculty of Management, Law and Social Sciences, and Lecturer of Strategic Human Resource Management and Change Management at University of Bradford, UK. Her academic-research career started when she was appointed as co-investigator for several multi-million pounds EU-funded research projects while completing her PhD in Brunel University, London. Prior to that, she worked in various organizations overseas, and her last role was a senior executive at one of the leading government agencies in Malaysia, where she had facilitated to spearhead the agency's human resources transformation and strategize the 10-year talent development plan. She draws from more than 15 years of these experiences to combine leadership, innovation, and entrepreneurship to achieve both personal and institutional success.

Dr. S.B. Goyal has peerless inquisitiveness and enthusiasm to get abreast with the latest development in the IT field. He has excellent command over industry revolution (IR) 4.0 technologies, such as Big Data, Data Sciences, Artificial Intelligence, Block Chain, etc. He is the first one to introduce IR 4.0 including Blockchain technology in the curriculum in Malaysia. He had participated as a speaker in the Bloconomic 2019 event on Blockchain. He had participated in many panel discussions on IR 4.0 technologies in academia as well as industry platforms. He has more than 18 years' experience in academia at national and international level. Currently, Dr. Goyal is associated as a Director-IT with City University, Malaysia.

Contributors

Nitesh Kumar Adichwal
University of Petroleum and Energy Studies (UPES)
Dehradun, India

Nitin Aggrawal
Department of Public Enterprises
New Delhi, India

Anni Arnav
School of Management
Presidency University
Bengaluru, India

Prakash Chandra Bahuguna
School of Business
University of Petroleum and Energy Studies
Dehradun, India

Deepak Bangwal
School of Business
University of Petroleum and Energy Studies
Dehradun, India

Manish Mohan Baral
Department of Operations
GITAM School of Business, GITAM (Deemed to be University)
Visakhapatnam, India

Pradeep Bedi
Department of Computer Science & Engineering
Galgotias University
Greater Noida, India

Chandan
Department of Professional Studies
CHRIST (Deemed-to-be University)
Bengaluru, India

Akhil Damodaran
School of Business
University of Petroleum and Energy Studies
Dehradun, India

Normala S. Govindarajo
School of Economics and Business Management
Xiamen University Malaysia
Sepang, Malaysia

S.B. Goyal
Faculty of Information Technology
City University
Petaling Jaya, Malaysia

Babita Jha
Faculty of Information Technology
School of Business and Management
Christ (Deemed-to-be University)
Bengaluru, India

Shankar Kumar Jha
Faculty of Information Technology
Management Department
SR Group of Institution
Jhansi, India

Dileep M. Kumar
Deputy Vice Chancellor (DVC) & Professor Strategy
Nile University of Nigeria
Abuja, Nigeria

Mukesh Kumar
Business Management
GNS University
Jamuhar, India

Rupesh Kumar
School of Business
University of Petroleum and Energy Studies
Dehradun, India

Najul Laskar
School of Business
University of Petroleum and Energy Studies
Dehradun, India

Pratibha Maurya
Shaheed Sukhdev College of Business Studies
University of Delhi
Delhi, India

Subhodeep Mukherjee
Department of Operations
GITAM School of Business, GITAM (Deemed to be University)
Visakhapatnam, India

Ipseeta Nanda
Faculty of Information Technology
GNS University
Jamuhar, India

Vijay Nimbalkar
Department of Management
International Institute of Management Studies
Pune, India

Surya Kant Pal
Department of Mathematics, School of Basic Sciences and Research
Sharda University
Greater Noida
India

Nishi Pathak
Department of Management Studies
Raj Kumar Goel Institute of Technology
Ghaziabad, India

Adith Potadar
Max Kelsen
Spring Hill, Australia

Lakshmi C. Radhakrishnan
Department of Management
Institute of Management &
Technology (IMT Business School)
Dubai, UAE

Yatika Rastogi
Department of Management Studies
Raj Kumar Goel Institute of Technology
Ghaziabad, India

Mohd Salim
Al-Tareeka Management Studies (ATMS)
Sharjah, UAE

Sunil Saxena
Microsoft
Sacramento, California, USA

Prayas Sharma
Management Department
Indian Institute of Management, Sirmaur (IIM)
Paonta Sahib, India

Sameer Shekhar
CTFL, Logistics Division, MoC
Indian Institute of Foreign Trade (IIFT)
New Delhi, India

Anurag Singh
School of Business
University of Petroleum and Energy Studies (UPES)
Dehradun, India

Ashish Kumar Singh
Department of Management Studies
Raj Kumar Goel Institute of Technology
Ghaziabad, India

Deepmala Singh
Symbiosis International University
SCMS
Nagpur, India

Rajeev Srivastava
Business Analytics Program, Decision Sciences
University of Petroleum and Energy Studies
Dehradun, India

Vishal Srivastava
School of Commerce
JAIN (Deemed-to-be University)
Bengaluru, India

Furquan Uddin
Department of Management and Business Administration
Aliah University
Kolkata, India

K. R. Varsha
Management
Dayananda Sagar University
Bengaluru, India

Chittipaka Venkataiah
Department of Operations
GITAM School of Business
GITAM (Deemed to be University)
Visakhapatnam, India

Abhinav Verma
University of Petroleum and Energy Studies (UPES)
Dehradun, India

Jason Walker
Psychology Department
University Canada West
Vancouver, Canada

Chapter 1

An Introduction to Business Intelligence

Shankar Kumar Jha
SR Group of Institution

Babita Jha
Christ (Deemed-to-be University)

Contents

DOI: 10.4324/9781003184928-1

1.1 Introduction

Business intelligence (BI) is a high-tech procedure to analyze information along with display applied and relates to a machine-readable storage medium that helps business professionals, enterprise leaders as well as consumers to improve their knowledge and understanding of corporation economic decision-making. In other words, BI software signifies the term computing, operations, as well as implementation with the aim of gathering, combination, and interpretation, including awareness about enterprise datasets. It is a group of applied science or computing that supports discovery, sloop, as well as analyzing leading information of significant numbers for better judgment or decisions. BI is an assimilation of plans, techniques, procedures, and structural frameworks, along with automation or robotics, which alter unprocessed data significantly, including beneficial information, with the aim of enterprise goals. It is the technique that includes procedures for improvement over enterprise economic decision-making in utilizing the intelligence of various references as well as carrying out information, including acceptance for improvement and correct insights into driving enterprises.

BI involves collecting, managing, as well as interpreting business data for generating information particularly shared by individuals all over the company and business for the betterment of complicated and unplanned as well as strategic judgment. BI constitutes a variety of mechanisms, usage, as well as program/scheme that allows corporations to collect, organize, monitor, and display relevant information for the business system.

1.2 Concept

BI software implementation supports businesses to get hold of data-driven analyzing promptly with more effective use as well as deliver information drawing out internally and externally for businesses. Consequently, BI is an evolving category of computerized usage since it utilizes data inventories to promote improvement in the adoption of the resolution. Various mechanisms, as well as methodologies of BI, such as data mining, data visualization, and predictive analysis, can be used to facilitate helpful information. BI software plays a major role in the achievement and effectiveness of operations, comprehensive schemes or plans, cooperation, as well as control. The application of BI is connected with the areas of selling, advertising, retailing, trading corporations, as well as customer support companies.

1.3 Literature Review

BI is an umbrella term (Levinson, 2006) and is structured to support the decision-making process (Gibson, Arnott & Jagielska, 2004). It is a relatively newly coined term by Howard Dresser in the early 1990s (Watson & Wixom, 2007). BI generally refers to a broad collection of software platforms, applications, and technologies that helps decision-makers to take an effective and efficient decision (Gibson, Arnott & Jagielska, 2004). It aids in taking better decisions (Clark, Jone & Armstrong, 2007) by facilitating the extraction of data and manipulating these data into information useful for making better business decisions (Bagale et al., 2021). BI enables to gain an understanding of capabilities available in an organization, future trends, or directions in the markets and technologies, competitors' actions, and the operating regulatory environment (Negash, 2004). The integration of operational data with analytical tools helps in taking timely and rational decisions (Kautish, 2008, Kautish & Thapliyal, 2013). BI system consists of tools, programmed products, and technologies that enable the collection, aggregation, integration, and quick availability of data (Yeoh & Koronios, 2010). The BI system helps in providing actionable information at the right time (Negash, 2004).

BI system consists of hardware solutions along with expensive software, on one hand, and specialized software, on the other (Olszak & Ziemba, 2006). The study further revealed that the requirement of a particular BI system depends on the needs of the business; however, any BI system requires a minimum of four specific components: Data warehouses, Extract, Transform, and Load (ETL) Tools, Online Analytical Processing (OLAP) Techniques, and Data Mining. The components of the BI system help in facilitating managerial decision-making and taking necessary actions (Golfarelli, Rizzi & Cella, 2004). Actions can further be described as acquiring information mainly with the support of the data warehouse component; gathering data with the help of extract–transform–load component; analyzing data with the support of online analytical products, and reporting of data supported by the data-mining components (Olszak & Ziemba, 2007). A critical factor on which the success of BI depends in any organization is the ability of an organization to take benefit of the available information for making an efficient decision (Cody, Kreulin, Krishna & Spangler, 2002). BI system is mainly used by organizations for proper management; monitoring of business-related activities; and planning, reporting, and decision-making, as well as maintaining customer relationships

(Olszak & Ziemba, 2007). More than ever, information supports all critical business decisions and BI seeks to provide the capability to access and analyze information (Matei, 2010).

BI systems can be used by different managerial levels for monitoring and improving strategic, tactical, and operational decisions (Stefan, Duica, Coman & Radu, 2010). BI facilitates senior-level management in getting inputs required for making strategic and tactical decisions, and at the lower managerial levels, it assists individuals in doing their routine jobs (Negash, 2004). At the strategic level, BI systems generate information that can be used for forecasting future results; at the tactical level, BI systems assist in decision-making for improving the company's performance; and at the operational level, they provide departmental performance based on just-in-time analysis (Olszak & Rziemba, 2007). Administrative choices of adopting BI are affected by enterprise information requirement, recovery, examination, and clarification (Nofal & Yusof, 2013). BI is the most preferred technology by CIOs because of its ability to facilitate improved decision-making (Watson & Wixom, 2007). Although BI systems are used widely by organizations, very limited research work has been done in this field (Negash, 2004). Numerous examples of BI can be witnessed that demonstrate the role of BI in overcoming adversity, but very little work has been done in research highlighting BI abilities (Schlegel & Sood, 2007). BI comprises numerical and methodological models for extracting data and valuable information from the available raw information. BI broadly consists of application, technologies, and processes for collecting, assimilating, accessing, and analyzing data that help users to make better decisions (Wixom & Watson, 2010). BI application in an organization enables the upgradation of knowledge by utilizing information through simulations and modeling. BI enables administrators in breaking down information from different sources to better leadership at both tactical and strategic levels (Rasoul & Mohammad, 2016). BI is a framework that transforms information into data and afterward into learning, consequently enhancing the company's basic decision-making process (Singh & Samalia, 2014). BI is categorized as a framework that collects, changes, and presents organized information from different sources lessening the required time to obtain noteworthy business data and enable their efficient use in management decision-making processes (Hamer, 2004). An organization that has made progress with their BI usage has stated that their BI is steady with their corporate business targets and much research on BI achievement concentrates on the alignment among BI and business targets (McMurchy, 2008).

1.4 Purpose and Significance of BI

1.4.1 Purpose

The ultimate goal of BI is to support and facilitate a corporation's economic performance. It allows an organization to access information or facts particularly crucial for its accomplishment. BI is applicable in various domains comprising selling, advertising, retailing, trading, financial institutions, including other domains as well as divisions/wings. Collecting facts or information is quite straightforward in recent times with the use of smart technology. BI intends to evaluate and interpret the collection of information as well as the following retrieved data. The retrieved data represent further changes toward understanding. Utilization of the recently collected information makes better enterprise. The objective of BI in enterprise is to help corporate executives, corporate leaders, as well as operational personnel to improve, including enhancing knowledge in economic business decision-making (Singh, Singh & Karki, 2021). Corporations too utilize BI for reducing expenditure, recognizing recent corporation's perspectives as well as marking ineffective corporation methods. BI comprises contributing information as well as management of the working mechanism, thereby facilitating organizations to increase their productivity. BI software regulates high-tech techniques, which comprise various connected operations and actions involving predictive analytics, benchmarking, text mining, process mining, prescriptive analysis, dashboard development, reporting, data visualization, OLAP, SQL, etc.

1.4.2 Significance

Every organization has to renew the datasets of consumer choices so that organizations are rapidly adjustable for altering requirements. BI software enables business executives to get timely information that aids them in taking the right decision at the right time. BI supports business executives as well as analysts to find out and meet the evolving new patterns. It supports and assists corporate executives as well as analysts in figuring out the adjustments to meet the evolving new patterns. BI enables the enterprises to go along with enhanced reasonable and functional information, facilitate better understanding of business patterns, and provide better business plan-furnished adoption of resolution replica.

BI is significant to understand the increase in current capacity as well as interpreting recent consumer-purchasing patterns. BI software applications enable business growth and development by enhancing the customer base. It helps and supports the organization's seller/salesman by providing virtualization of data of buying process and aids in investigating overall income assessment. Additionally, BI promotes seller/salesperson in recognizing functions involving detecting the cause of the problems, thereby providing outcomes immediately to improve sales. Besides, BI is utilized for the development of commodity/byproducts to meet their expectations and needs. Also, it enables the collection of facts for generating byproducts for enterprises. It helps in increasing income for those enterprises that have low productivity. Generally, selling and retailing personnel observe their consumers or clients for monitoring or trailing their needs, and for this purpose they use customer relationship management (CRM) technology. CRM facilitates end-users to access databases online to predict the needs of consumers with the use of the operating system, virtual fascia, as well as routine chat messages or electronic messages.

BI allows enterprises in contributing facts concerning the present action database during computerized or automated information extraction, as well as comprehending facts. Most appropriate techniques to deliver datasets involve better-improved methods that allow companies to remove unnecessary functions and responsibilities, enabling personnel to highlight their functions rather than be concerned about the information-handling process. The personnel/HR division is also adopting BI software for productiveness, interpretation, income trailing, payment and remuneration, and perception toward personnel happiness. Fiscal reports are more crucial for any organization in the finance division for better understanding as well as for better insights. Moreover, BI supports for periodical reports, which include yearly balance sheets and recognize future difficulties (Kautish & Thapliyal, 2013). BI allows companies better forecast revenue, cost-effective cutting, as well as increase sales, including market expansion, and also provide chances for the continuous development of enterprises or entrepreneurship (Kautish, Singh, Polkowski, Mayura & Jeyanthi, 2021). Finally, BI provides feasibility to amalgamate datasets through various references or resources, interpretation of database toward the layout as well as spread it for the intelligence of pertinent team members. Presently, it enables companies to have an overall approach as well as create smart line judgment.

1.5 Features of BI

Organizations have a huge volume of data and managing these data becomes a tedious task. BI tools help in easy accessibility of data and this leads to better decision-making in an organization. Several features of BI make it user-friendly and easy to operate. The features associated with BI are categorized into High-level, Safety, and Essential.

1.5.1 High-Level Features

1.5.1.1 Management of Metadata

The crucial feature of BI is Metadata management. Metadata management denotes activities involved in the efficient management of data and outcomes related to it. It is mainly categorized into three activities, namely technical, business, and operational. Metadata management includes policies, processes, responsibilities, and roles that ensure that data-based information is obtainable, reachable, supportable, and sharable across an organization. Metadata management mainly concentrates on indicators, organization, measures, and other aspects of data desired for business analysis. The use of BI in metadata ensures the quality, completeness, and consistency of the data in use.

1.5.1.2 Data Visualization

In today's technological era, data visualization has become an important aspect of BI. Data visualization helps in presenting bulk data in a more simplified and significant manner. Data visualization through BI has more advantages in comparison to traditional visualization (Tunowski, 2015). Data can be significantly visualized with the help of conventional charts and special charts. Special charts refer to visual special effects, such as flow maps, heat maps, rectangular treemaps, etc. This enables us to visualize complex data and understand it at a glance. BI-generated conventional and special charts help in properly exploring and analyzing data.

1.5.1.3 Analytics Dashboards

Dashboards have become an important feature of the BI platform because they help in displaying, analyzing, comparing, customizing, and sharing data. These dashboards help in the management of information and also provide data visualization solutions. Snapshot overviews can be generated by applying filters along with interactive chart components. Key performance indicators (KPIs) can be revealed with the help of analytics dashboards. Companies can achieve greater success by monitoring their dashboards regularly.

1.5.1.4 Mobile Support

Mobile BI helps anytime and anywhere in accessing data. It helps in accessing BI data, such as KPIs, customized dashboards, and business metrics on mobile devices. The users of BI software can easily access, sort, administer, and visualize relevant data at any time. According to the Mobile BI Market Report (2020), the mobile BI market size is forecasted to grow at a compounded annual growth rate of 22.2% from 2016 to 2021.

1.5.1.5 Real-Time Data

Real-time data can be easily and timely delivered using BI tools. Real-time data help in gaining full knowledge and facilitate in making the best decisions. The real-time feature of a business system enables an enterprise to store data related to business transactions as and when they occur. It also supports in taking quick actions as real-time data are available for making a strategic decision.

1.5.2 Safety Features

1.5.2.1 Alternative Authentication Sources

Advanced BI software provides alternative authentication sources. With the help of alternative authentication sources, a manager can use the existing authentication method and can also add alternative authentication sources for different internal systems.

1.5.2.2 Security at the Application Level

Application access in BI is usually based on the needs and preferences of the user. Based on the requirements of a particular business, effective filters can be applied and thereby internal security can be tightened. Effective security helps in ensuring that only relevant business applications go into the hands of the members of the organization.

1.5.2.3 Row-Level Security

Row-level security allows data to be safely stored for multiple users. The security policy of row-level allows filtering a row that does not belong to a particular user/tenant. Row-level or multi-tenant security mitigates security errors.

1.5.2.4 Activity Auditing

Activity auditing allows operators to monitor the activity of an end-user after each session. This feature gives an assurance to the end-user that the IT operator can check the log-in time, applications accessed, and time duration of each session.

1.5.3 Essential Features

1.5.3.1 Ad Hoc Reports

Ad hoc reports are one of the important features of BI that facilitate business developers to generate reports on the go. Relevant data can be chosen as per the requirement and users can produce reports in their desired format. The generated report can be directly mailed to the concerned party from the Web.

1.5.3.2 Ranking Reports

This particular report option enables users to view the best and worst aspects of a particular business. This feature in the BI solution helps in creating ranks across multiple business dimensions. Various selection criteria can be taken into consideration and ranking reports can be prepared accordingly.

1.5.3.3 Executive Dashboards

Personalized Dashboards give business leaders real-time data in the form of summaries, graphs, charts, etc. The availability of relevant and easy-to-understand data helps organization leaders to make better decisions and increases the organization's effectiveness. Executive Dashboards by displaying the KPIs of an organization help in the smooth running of an organization.

1.5.3.4 Pivot Tables

Pivot tables are significant tools that provide views of multidimensional data in a tabular form. They extract substantial data from large, messy data and enable data consolidation, comparison, and summarization. Pivot tables, on the one hand, aid in calculations, such as counting, sorting, and averaging data in one table, and, on the other hand, present summarized reports on the other table. Pivot tables are important tools for examining information and predicting trends.

1.6 Importance of BI

1.6.1 Fast and Precise Reporting

BI systems help in the faster understanding of data and therefore enables quicker and faster decisions in an organization (Luminita & Magdalena, 2012). The customized reports generated through BI software enable people of an organization to monitor their KPIs using data sources related to production, sales, and finance. Real-time reports can be produced and accordingly actions can be quickly taken. Visualizations, such as charts, graphs, and tables, make reports easy to read.

1.6.2 Better Data Quality

Data generated in a business are bulk in nature. These data need to be organized and presented effectively. BI software helps in collecting and generating quality data. It aids in combining various data sources, thereby enabling management to get a clear picture of their business.

1.6.3 Ascertain Market Trends

In this competitive era, businesses need supportive data to identify new opportunities and formulate new strategies to enhance their profitability. Organizations can have access to external data and can relate that data with their internal data to detect new market trends and market conditions.

1.6.4 Enhanced Customer Satisfaction

Customer feedback is taken by most companies in real-time, and this helps companies in retaining and reaching new customers. BI software provides tools that help in identifying purchasing behavior of the customers. With the help of available tools of BI, companies can determine customer needs and deliver better services.

1.6.5 Increased Functional Efficiency

Multiple data sources can be unified with the support of BI tools, and this helps in reducing the tracking time of relevant information by managers and employees. BI software produces accurate and timely reports that help employees to focus on their goals.

1.6.6 Gain Business Insights

Employee productivity and department-specific performance can easily be gauged with the use of BI tools. Businesses can know their strengths and weaknesses. Alerts can easily be installed and performance metrics can be traced through them.

1.6.7 Competitive Analysis

BI software helps in planning, budgeting, and forecasting data. Organizations can analyze their data and can easily check competitors' performance. Competitive strategies can be designed by taking into consideration the available information.

1.6.8 Real-Time Data

BI system delivers real-time data that can be accessed through dashboards, worksheets, and scheduled emails. BI tools aid in the assimilation, interpretation, and distribution of real-time data.

1.6.9 Improved Visibility

Visibility of the functions of a business can be enhanced with the use of an improved BI system. Organizations can examine the areas for improvement and can have better control over their processes.

1.6.10 Data-Driven Business Decision

Better business decisions can be taken because of the availability of timely and accurate data. Nowadays, customized mobile dashboards are available for the functional team, which helps in the availability of real-time data.

1.6.11 Enhances Employee Satisfaction

BI software helps employees to analyze data with little training. Employee satisfaction increases because they have to rely less on IT support for the data. BI software is designed in such a manner that it provides data solutions to various departments based on their requirements.

1.7 BI Architecture

"A BI Architecture (BIA) is the framework for the various technologies an organization deploys to run BI and analytics application" (Pratt & Fruhlinger, 2020). BIA comprises the data-processing system as well as information technology programs that are utilized for gathering and combining storage along with investigating BI databases. BIA is a central and essential component in the implementation of effective and efficient BI tools, which utilizes analysis of information with reports to support business follow-up commercial operations, recognize advanced income scope, better tactical planning, including improving better appropriate action/solution of crisis management throughout.

1.7.1 Architecture of BI

Vercillis (2009) utilizes hierarchy to illustrate through which BI structure is built as follows:

1.7.1.1 Data Sources

The sources primarily comprise databases connected with operationalizing methods, even so possibly also involve unorganized or unstructured data, such as electronic mail and information, acquired deriving out of outsourcing.

1.7.1.2 Data Warehouse/Data Mart

Data warehouses are utilized to combine various types of databases into a central location utilizing a method called ETL as well as standardize these consequences throughout the structure, which enable prospective questions.

Data Marts are normally small warehouses that emphasize core databases of a sole department, alternatively gathering information throughout an organization. They confine the complication of information as well as inexpensive to implementing than full warehouses.

1.7.1.3 Data Exploration

It is an inactive or passive BI investigation comprising the question as well as reported systems and mathematical methods.

1.7.1.4 Data Mining

It is a dynamic and effective (or active) BI procedure and technique along with the motive of the database as well as knowledge extraction deriving out of the database.

1.7.1.5 Optimization

This model enables to find out a better alternative from a class or group of different activities, which is generally quite large as well as periodically still indefinite.

1.7.1.6 Decisions

Although BI procedures and techniques are accessible as well as effectively and efficiently embraced, the option of decision-making relates to an administrator, who can take advantage of non-formal with non-structured intelligence accessible to adjust as well as modify the suggestions, including the consequences accomplished over the utilization of statistical methods.

1.8 BI Tools

There are various BI tools, which is beneficial for the achievement of digital objectives. It also supports gaining knowledge and insights and solving distinct problems facing everyday databases. Some of the recent BI tools available are as follows:

1.8.1 Data Pine

It is a multipurpose BI platform that facilitates complicated procedures about the analysis of data without a technical end-user. Its remedy allows analysis of data as well as corporate customers identical to readily assimilate various sources of information, carry out sources of information, carry out modern analysis of data, make connected enterprise scoreboard, or control panel as well as create business-relevant information.

1.8.2 SAP Business Intelligence

It provides modern data analysis remedy as well as online BI, prediction or prospective analysis, expert systems or robotics, as well as tactical analysis or strategic thinking, including analysis of information. This BI Platform especially provides submission of data as well as data evaluation, visualizes datasets, investigates utilization, workplace assimilation, including analysis of portability. SAP BI is a hale computing plan for each specific responsibility and function, for instance computational, customers, and administration as well as workplace multitude of office bearers in a single framework.

1.8.3 Yellowfin BI

This BI tool is used across the country, i.e., end-to-end analytics platform, which integrates visual images, natural language processing (NLP), and a combination of processing techniques. It readily cleanses with plenty of information and emotional cleansing, for instance, tab and button related, and switches along with opening up scoreboard or control panel virtually everywhere.

1.8.4 Micro Strategy

This BI tool provides a robust, effective, and broadband, control panel as well as analyzes the data that support control trends, identify current chances, better productiveness as well as many more. End-users are possibly linked to each and various references and either the inward information is from worksheets, cloud computing, or enterprise application software. It can be retrieved from of a personal computer or through a portable.

1.8.5 SAS Business Intelligence

Although SAS is more prevalent providing modern forward-looking analytics, such as predictive analytics, it also facilitates an important BI platform.

It is a self-serving mechanism that enables leveraging information as well as measuring to create appropriate solution concerning their enterprises. Utilizing a group of Application Programming Interfaces facilitates a lot of personalization choices. SAS ensures prominent data automation as well as modern data mining or data modeling, including data reporting.

1.8.6 Zoho Analytics

This BI tool is utilized for the comprehensive submission of data as well as for the analysis of data. It has automatic or computing data-processing synchronization and possibly be regular time or standard time. It can readily make a connecter by utilizing the combination of Application Programming Interfaces and mix and make information from various references as well as make purposeful information. It makes it simple to interpret or report while generating customization reports as well as a screen board or dashboard allowing to zoom in for vital descriptions.

1.8.7 Oracle BI

It is a business notebook for computing as well as usage for BI. It provides end-users with all reasonable BI competencies, such as control panel (dashboard), enterprising intellect and understanding, carefulness, makeshift, and much more. It is too better for an organization that requires interpretation of big size database deriving out of Oracle as well as sources of non-Oracle in reality, which is an extremely healthy remedy.

1.8.8 Looker

Looker is a BI tool for a data discovery application to keep an eye throughout. BI tool platform combines with each structured Query Language dataset and repository. Looker is excellent for startups, medium-sized enterprises, and business-class enterprises. Several advantages of this specific device comprise user-friendliness, useful and helpful envision, close collaboration, and information sharing through electronic mail or USL and are combined with other usages, including staunch help from the technical staff.

1.8.9 Microsoft Power BI

It is based on an online business architecture analyst tool suite that excels in database visualization. It enables end-users to recognize patterns in actual time as well as marked on modern connectors, which enable raising the game in motion. Due to its online-based, the MicroSoft power is possibly accessible driving out of reasonably everywhere. It is a kind of software that enables end-users to a combination of their applications as well as report transfer, including an actual-time control panel.

1.8.10 Clear Analytics

It is an intuitive or emotional Excel-based software BI tool, which is utilized by personnel and also for great fundamental drive in BI network, which provides distinct BI characteristics, such as generating, robotic, survey analysis as well as visual perception of the database of an organization.

1.8.11 Domo

It is cloud computing predicated on the BI platform, which combines various sources of data as well as worksheets or Excel sheet datasets, including networking. It is utilized equally by small-scale enterprises as well as big multinationals. This BI platform provides micro as well as macro-level transparency as well as investigation. Beginning at cash flow situation with an inventory of the best selling products between zones for computations of the marketing investment returns for all mediums. This is a barely disappointing concern to Domo, which is complicated to upload interpretation deriving out of the cloud for its use.

1.8.12 Tableau

It is a BI software for the exploration of datasets as well as visualization of data. By using this software one can readily interpret, visualize, as well as contribute information excluding computing to interfere. It promotes various sources of information, such as Google analytics, salesforce, MS Excel, Oracle, and SQL. One of the best things about Tableau is there is no charge for using it. It provides gives users highly developed control panels that are extremely accessible. Besides, it also provides self-sufficient products, such as Table Desktop, for everyone, as well as Tableau server analytics for organizations that can compete globally.

1.8.13 IBM Cognoscenti Analytics

It is an Artificial Intelligence-sustained BI platform that promotes the whole assessment cycle deriving out of finding for implementation. One can interpret, visualize, as well as exchange relevant information with regard to information among teammates. This is the biggest advantage of Artificial Intelligence, which one can compete to find out the concealed trends, due to the information as per the facts analyzed as well as displayed in the viewing report.

1.8.14 Microsoft Power BI

This BI is based on an online business architecture process mechanism suite that is excellent in visualizing datasets. This enables end-users to recognize patterns in real-time as well as mark modern connectors, which enables raising the game in motion. Due to its online-based, MS Power BI can be accessed reasonably everywhere. This kind of program enables end-users to combine their applications as well as transfer reports, including real-time control panels.

1.8.15 Sisense

This BI tool is an accessible mechanism that enables everyone inside the company to control major information in a complicated database, including interpretation as well as conceptualizing this information in the absence of computer engineering.

It reunites information deriving out of a broad range of sources and comprising Salesforce, AdWords, and Google analytics. Besides, due to its use in open-end credit, processing of data is relatively against other devices.

1.8.16 Qlik Sense

This BI tool for an organization is a qlick of a product known as Qlik View. It is also utilized for sense for qlik driving out of whatever tool all the time. Most end-users connect for Qlik Sense, which is home for touchpad making and it a highly successful BI tool. The major distinction of Qlik View is perspective storytelling. End-users sum up a distinctive understanding of the information as well as during utilizing photographs/snapshots, including a focus on making a correct interpretation as well as understanding, which is now much simpler.

1.9 Components of BI

BI includes five components. They are as follows:

1.9.1 Online Analytical Processing

It is an important component for the success of the modern organization that can take advantage of all available information (Cody et al., 2002). OLAP refers to the techniques of performing complex analysis over the information stored in a data warehouse to transfer it into decision information. It is a computing method that allows users to sort, select, and analyze data for strategic decisions. It has become a vital tool for analysts, managers, and executives for analyzing and extracting various interesting patterns from a large volume of stored data (Techapichetvanich & Datta, 2005). It helps in financial reporting, market forecasting, budgeting, trend analysis, and other planning purposes. OLAP systems can be categorized into Multidimensional OLAP, Relational OLAP, and Hybrid OLAP. OLAP process starts with the accumulation of data from various sources and these accumulated data are then collected in a data warehouse. The data stored in the warehouse are then cleansed and are accumulated in OLAP Cubes from which users can generate queries.

1.9.2 Corporate Performance Management

Corporate performance management (CPM) is an advanced analytical tool that converts the relevant information into operational plans. It comprises activities, metrics, methods, and processes, which are used in managing and monitoring the performance of a business. Organizations can have an integrated view of their planning and forecasting. CPM software helps in cost reduction, budget remodeling, proper alignment of KPIs, and improving financial planning.

1.9.3 Real-Time BI

In recent years, the Real-Time component of BI has been gaining popularity. It aids in disseminating real-time data through electronic mails, interactive portals, and messaging systems (Ranjan, 2005). Organizations can have access to the most recent and relevant data. With the help of graphs, charts, etc., visualization of data is also possible. A business can easily respond to a particular situation because of the availability of real data. For instance, if a financial institution wants to decide on extending credit to a particular party, then real-time credit scoring helps in taking this decision. Real-time data help in tracking customer data, enable real-time testing, enhance response time, and reduce cost.

1.9.4 Data Warehousing

Knowledge workers can be empowered with information through data warehousing (Herschel & Jones, 2005). Data warehousing allows the owners of the business to go through different subsets of the data and analyze various components of the data to make the right business decisions. It is the core component of the BI system that enables analyzing and reporting of data. Businesses and industries can generate important statistics with the help of Data Warehousing. The common sectors using Data Warehousing include airlines, banking, healthcare, insurance, retail chain, telecommunications, etc. It helps analysts in integrating data warehouse applications with BI tools, such as Chartio, Looker, Tableau, etc. Data warehousing can be used to empower knowledge workers with information that allows them to make decisions based on a solid foundation of fact.

Data warehouses have four components: Load Manager, Warehouse Manager, Query Manager, and End-User Access tools. Data warehouse receives data from apps and systems used in an organization, and these data are arranged in a particular format. After formatting, these data are imported to match with the already stored data. The processed data are then available for the user to make relevant business decisions. With the help of Data Warehouse, turnaround time for analyzing and reporting is reduced.

1.9.5 Data Sources

The data in use originate from a location known as the data source. A huge amount of data are generated in companies for operational purposes. Proper understanding of data sources is important for the integration of the BI process and methodology. Data sources are navigated by the BI system. For various analytical options, raw data from internal and external sources have to be pulled out. Companies tend to store huge amounts of operational data. Organizations can take advantage of the following operational data sources:

1.9.6 Enterprise Resource Planning

It is a business management software that is used by the organization for collecting, managing, integrating, and interpreting data. It is used by large companies and includes modules related to financials, supply chain, manufacturing, human resource activities, etc.

1.9.7 Customer Relationship Management

CRM software helps in understanding and analyzing customer behavior, improves customer relationships, and aids in making better business decisions.

1.9.8 E-Commerce Apps

These apps help in sourcing data that can lead to real-time sales activity. It helps in better understanding of buying behavior among different segments of the customer and also helps inventory management of e-commerce companies.

Other sources are numerous company databases, flat files, web services (apps), RSS feeds, and more.

1.10 BI Limitations and Challenges

1.10.1 Threat of Security Breach

In today's technological era, the biggest challenge businesses are facing is the security issues with any data analysis system. The risk of data hacking and data leakage is always present with most of the analysis systems. An error in the BI process can expose the sensitive information stored in it. Data hackers can destroy, steal, or even stop the authorized user from accessing a particular BI system. Data breaching is the biggest threat associated with BI tools.

1.10.2 High Prices

BI software mostly proves expensive for small and medium-sized companies. Software along with the cost of the hardware makes it more expensive. There are several vendors available for a particular software but there are variations in the services provided and prices are also not transparent. Data management also becomes costlier for companies using mobile or cloud-based solutions for their BI applications.

1.10.3 Difficulty in Examining Various Data Sources

BI system allows you to go through various data sources at a time. Using different data sources proves beneficial for analytics, but the access of data through multiple platforms may prove troublesome for the system. However, this problem is being resolved in the more advanced BI system. They are offering services in the form of independent tools that allows the consolidation of varied information.

1.10.4 Unwillingness to Adoption

There may be resistance from a particular department or employees to integrate their operation with BI software. Adequate training is required for the implementation of the software. Unwillingness to learn or adopt new software may become a hindrance in the proper operation of the BI system.

1.10.5 Blending of Professional and Personal Information

Employees may be required to use their mobile or personal devices to access the information stored in the BI system. Since the personal data are also stored in personal devices there may be instances of the blending of professional and personal data.

1.10.6 Need for Multiple BI Applications

Multiple BI applications are available today. Most of the BI systems provide one specific surface. A fully BI integrated system is required to effectively analyze big data and, for this, an organization may be required to invest in an umbrella of services.

1.10.7 Complexity

Another limitation of a BI system is the complexity in data implementation. Sometimes the implementation of data becomes so complicated that it leads to rigidity in the business techniques.

1.11 Limitations with BI Dashboards

With the incorporation of huge data, there may be a possibility of flashy or cluttered design in a dashboard. This may create difficulty in examining relevant information. Data cannot be refreshed automatically on a dashboard; it has to be done manually. There is also difficulty in attaching supporting data.

1.11.1 Different Conclusion from the Same Data

BI provides consistent analytics. These data may be inferred differently by different individuals. Due to conflicts in inferences, there may be a delay in taking decisions.

1.12 Pathway to BI Success

BI success in an organization is not automatic and it depends on the conditions facilitating it. The following major factors should be taken into consideration for BI success.

1.12.1 Senior Management Support

For the success of the BI system, top management should believe and drive the application of BI in an organization. BI effectiveness can be increased through strong and committed leadership (Seah, Hsieh & Weng, 2010). The important factor for BI system implementation is consistent management support and sponsorship (Yeoh & Koronios, 2010). Top management should have a clear vision for the use of a particular BI solution, and decisions should mainly be taken based on information analyzed through BI tools. They should provide all the necessary resources and support for the proper implementation of BI solutions.

1.12.2 Effective Validation Process

Several data are uncovered by using BI software, but only valid data have to be taken into consideration for decision making. BI system should be structured in such a manner that it prevents problematic data from entering a system. This will help in getting a true insight into particular information. BI system should be designed in such a manner that all the data required for answering different queries can be accessed easily.

1.12.3 Regular Monitoring and Adjustment of BI Use

Regular monitoring of the BI system should be done by the IT department. This will help in getting a clear picture of the data sources being accessed, tools being used, and the manner or pattern of using a particular tool in an organization.

1.12.4 Skilled or Qualified Staff

Human assets play a vital role in the successful implementation of the BI (Bozic & Dimovski, 2019). Adequate knowledge, experience, and expertise of the project members are required for the successful implementation of BI solutions (Arnott, 2008). The experiences, knowledge, and meanings of the members of a team play an important role in BI success (Villamarin & Diaz, 2017).

1.12.5 Development Environment

The development environment is a critical success factor that is mostly ignored by organizations. The components of the development environment encompass development methodologies, development tools, project management teams, etc. An organization must focus on the components of the development environment for the implementation of a BI system.

1.12.6 Proper Selection of Technology

Organizations should properly evaluate different BI options. The selection of BI solutions should be based on the needs, required skills, and budgets of an organization. Proper selection of technology will help in solving decision issues in an organization.

1.13 Future of BI

1.13.1 Data Storytelling

Data storytelling will become a norm in the coming years. Storytelling is a method through which insightful information can be communicated in narrative form. The application of BI in an organization mostly helps in providing insights into particular data. However, if data are conveyed in an actionable and reasonable manner, it will surely help in taking rational and fast decisions. Storytelling will provide narratives that will help in a better understanding of data in use.

1.13.2 Data Governance

In the growing era of BI, data governance will become a priority for small and big organizations. Data governance allows the management of data in such a manner that complete, reliable, and secure data are delivered. A comprehensive data governance strategy helps in improving return on investment from the investment made in BI software. It also aids in maintaining a sound balance between the consistency and transparency of data. Data governance will help in getting access to accurate data that will help in making the right decision. It will also ensure the privacy and confidentiality of the data procured and will also prevent its use by unauthorized users.

1.13.3 Self-Service BI

Although self-service BI is in existence, it will become a necessity in the coming years. Business users expect that data can be accessed anytime and anywhere so that all the tasks can be performed smoothly. Without relying much on the analytics team, business users want to solve their problems on their own. They want to use the tools of their choice to drive value. It will enable business users to become self-sufficient and will break the reliability of IT teams for the access of the right data. It will help in making analysis easier and crucial decisions can be taken at a faster pace.

1.13.4 Prescriptive Analysis

Prescriptive analysis helps in finding the best solution from different available alternatives. It is linked to descriptive and predictive analytics, but it focuses more on actionable insights. The prescriptive analysis will become a worldwide feature in the coming years. Each decision will be gauged from the future perspective and thereby it will help in taking the best decisions. Prescriptive analytics will be used by most organizations in the future because it takes into consideration all possible scenarios, past performance records, available resources, etc. and suggests the best possible action or solution.

1.13.5 Collaborative BI

Integration of all the relevant information related to decision-making (Azeroual & Theel, 2018) will be an important feature of the BI. Collaborative BI implies the assimilation of BI and analytical capability, with the organization's tools and techniques to share information, collaborate, and exchange perspectives for business information analysis. Presently, BI tools are mostly operated independently and are not linked to a broader network. The next-generation BI system will witness more business users of

Collaborative BI. With the application of Collaborative BI, enhanced knowledge sharing will be there and this will lead to an increase in productivity in an organization.

1.13.6 Natural Language Processing in BI

NLP assists in revealing patterns from the unstructured data making it more useful for further analysis. NLP will be the future of BI. With growth in technology, NLP applications will become more user-oriented and will help in easy accessibility of information from a complex database. Big brands are already using the techniques of NLP for analyzing customer sentiments by extracting particular information from a piece of text. This is also known as opinion mining in market parlance. In the future, NLP will be used by all the business areas in an organization.

1.14 Conclusion

In the data-driven world, BI is becoming a buzzword. Nowadays, enterprises are paying serious attention toward the investment in the BI system to meet the needs and wants of the customer. BI has enabled organizations to take advantage of business opportunities and has increased their competitive edge. With the quick and timely availability of quality data, an organization can make an optimal decision at any point in time. Although the BI system has been adopted by many large corporations, some of them still find challenges in adopting it. These challenges can be overcome if proper attention is given to its successful implementation. The successful implementation of BI depends on leadership, quality management, organizational culture, human capital, regular monitoring, etc.

References

Arnott, D. (2008). Success factors for data warehouse and business intelligence systems. In *ACIS 2008 Proceedings -19th Australasian Conference on Information Systems*, 55–65.

Azeroual, O., & Theel, H. (2018). The effects of using business intelligence systems on an excellence management and decision-making process by start-up companies: A case study. *International Journal of Management Science and Business Administration*, 4(3), 30–40.

Bagale, G. S., Vandadi, V. R., Singh, D., et al. (2021). Small and medium-sized enterprises' contribution in digital technology. *Annals of Operations Research*. Doi: 10.1007/s10479-021-04235-5.

Bozic, K., & Dimovski, V. (2019). Business intelligence and analytics for value creation: The role of absorptive capacity. *International Journal of Information Management*, 46, 93–103.

Clark, T. D., Jones, M. C., & Armstrong, C. P. (2007). The dynamic structure of management support systems: Theory development, research focus, and direction. *MIS Quarterly*, 31(3), 579–615.

Cody, W. F., Kreulin, J. T., Krishna, V., & Spangler, W. S. (2002). The integration of business intelligence and knowledge management. *IBM Systems Journal*, 41(4), 697–713.

Gibson, M. C., Arnott, D. R., & Jagielska, I. (2004). Evaluating the intangible benefits of business intelligence: Review & research agenda. In R. Meredith, G. Shanks, D. Arnott, & S. Carlsson (Eds.), *Decision Support in an Uncertain World: Proceedings of the 2004 IFIP International Conference on Decision Support Systems (DSS2004)*, 295–305.

Golfarelli, M., Rizzi, S., & Cella, I. (2004). Beyond data warehousing. In *Proceedings of the 7th ACM International Workshop on Data Warehousing and OLAP - DOLAP '04*. Doi:10.1145/1031763.1031765.

Hamer, D. P. (2005). *The Organization of Business Intelligence*. The Hague: SDU Publishers.

Herschel, R. T., & Jones, N. E. (2005). Knowledge management and business intelligence: The importance of integration. *Journal of Knowledge Management*, 9(4), 45–55.

Levinson, M. (2006). *Business Intelligence: Not Just for Bosses Anymore*. CIO. Retrieved from http://www.cio.com/article/16544/Business_Intelligence_Not_Just_for_Bosses_Anymore.

Luminiţa, S., & Magdalena, R. (2012). Optimizing time in business with business intelligence solution. *Procedia - Social and Behavioral Sciences*, 62, 638–648.

Kautish, S. (2008). Online banking: A paradigm shift. *E-Business, ICFAI Publication, Hyderabad*, 9(10), 54–59.

Kautish, S., Singh, D., Polkowski, Z., Mayura, A., & Jeyanthi, M. *Knowledge Management and Web 3.0: Next Generation Business Models*. Berlin: De Gruyter.

Kautish, S., & Thapliyal, M. P. (2013). Design of new architecture for model management systems using knowledge sharing concept. *International Journal of Computer Applications*, 62(11).

Matei, G. (2010). A collaborative approach of business intelligence systems. *Journal of Applied Collaborative Systems*, 2(2), 91–101. Retrieved from http://www.jacs.ro/2010-Volume02/number02/paper009-fullpaper.pdf.

McMurchy, N. (2008). Survey of bi purchase drivers shows need for new approach to business intelligence, July 2008, Available online at http://www.gartner.com/id=714209.

Mobile Business Intelligence Market - Growth, Trends, Covid-19 impact, and Forecasts. (2021–2026). Retrieved from https://www.mordorintelligence.com/industry-reports/mobile-business-intelligence-market.

Negash, S. (2004). Business intelligence. *Communications of the Association for Information Systems*, 13. Doi: 10.17705/1CAIS.01315.

Nofal, M. I., & Yusof, Z. M. (2013). Integration of business intelligence and enterprise resource planning within organizations. *Procedia Technology*, 11, 658–665.

Olszak, C. M., & Ziemba, E. (2006). Business intelligence systems in the holistic infrastructure development supporting decision-making in organizations. *Interdisciplinary Journal of Information, Knowledge and Management*, 1, 47–58.

Olszak, C. M., & Ziemba, E. (2007). Approach to building and implementing business intelligence systems. *Interdisciplinary Journal of Information, Knowledge and Management*, 2, 135–148. Retrieved from http://www.ijikm.org/Volume2/IJIKMv2p135-148Olszak184.pdf.

Pratt, M. K., & Fruhlinger. (2019). What is business intelligence? Transforming data into business insights. CIO, India. Retrieved from https://www.cio.com/article/2439504/business-intelligence-definition-and-solutions.html.

Ranjan, J. (2005). Business intelligence: Concepts, components, techniques, and benefits. *Journal of Theoretical and Applied Information Technology*, 9(1), 600–607.

Rasoul, D. G., & Mohammad, H. (2016). A model of measuring the direct and impact of business intelligence on organizational agility with partial mediatory role of Empowerment: Tehran construction Engineering Organization (TCEO) and EKTA organization industries.co. *Social and Behavioral Sciences*, 230, 413–421.

Schlegel, K., & Sood, K. (2007) Business intelligence platform capability matrix. Retrieved April 01, 2008, from http://www.informationbuilders.com/product s/webfocus/pdf/Gartner_BI_Matrix.pdf.

Seah, M., Hsieh, M. H., & Weng, P.-D. (2010). A case analysis of Savecom: The role of indigenous leadership in implementing a business intelligence system. *International Journal of Information Management*, 30(4), 368–373.

Singh, D., Singh, A., & Karki, S. (2021). Knowledge management and Web 3.0: Introduction to future and challenges. In S. Kautish, D. Singh, Z. Polkowski, A. Mayura, and M. Jeyanthi (Eds.), *Knowledge Management and Web 3.0: Next Generation Business Models*, 1–14. Berlin: De Gruyter. Doi: 10.1515/9783110722789-001.

Singh, H., & Samalia, H. (2014). A business intelligence perspective for churn management. *Procedia – Social and Behavioral Sciences*, 109, 51–56. Doi:10.1016/j.sbspro.2013.12.420.

Stefan, V., Duica, M., Coman, M., & Radu, V. (2010). Enterprise performance management with business intelligence solution. In L. A. Zadeh et al. (Eds.), *Book Series Recent Advances in Computer Engineering: 4th WSEAS International Conference on Business Administration (ICBA'10)*, 244–255. Cambridge, England: University Cambridge.

Techapichetvanich, K., & Datta, A. (2005). Interactive visualization for OLAP. In: O. Gervasi et al. (Eds.), *Computational Science and Its Applications – ICCSA 2005. Lecture Notes in Computer Science*, vol 3482. Berlin, Heidelberg: Springer.

Tunowski, R. (2015). Business intelligence in organization. Benefits, risks and developments. *Entrepreneurship and Management*, 16(2), 133–144.

Vercillis, C. (2009). *Business Intelligence: Data Mining and Optimization for Decision Making*. Politecnico di Milano, Italy: John Wiley & Sons Ltd. Retrieved from https://onlinelibrary.wiley.com/doi/book/10.1002/9780470753866.

Villamarín García, J. M., & Díaz Pinzón, B. H. (2017) Key success factors to business intelligence solution implementation. *Journal of Intelligence Studies in Business*, 7(1), 48–69.

Watson, H. J., & Wixom, B. H. (2007). The current state of business intelligence. *Computer*, 40(9), 96–99. Doi:10.1109/mc.2007.331.

Wixom, B., & Watson, H. (2010). The BI-based organization. *International Journal of Business Intelligence Research (IJBIR)*, 1(1), 13–28. Doi: 10.4018/jbir.2010071702.

Yeoh, W., & Koronios, A. (2010). Critical success factors for business intelligence systems. *Journal of Computer Information Systems*, 50 (3), 23–32.

Chapter 2

Business Intelligence: A Value-Increasing Strategy for the Organization

Najul Laskar
University of Petroleum and Energy Studies

Deepmala Singh
Symbiosis International University

Contents

DOI: 10.4324/9781003184928-2

2.1 Introduction

Many business owners are being overwhelmed by enormous amounts of data in today's big data world. They are always looking for methods to improve their knowledge and control of this information in order to maximize the value for their organization. Business intelligence (BI) is a suite of techniques, tools, software, and best practices that assist businesses in collecting, integrating, analyzing, and presenting raw data into easily understandable, thoughtful, and enforceable business information to enhance and accelerate performances and decision-making. Many firms today face the issue of making strategic decisions in a fast-paced commercial climate. Although most firms excel at budgeting, reporting on financial and non-financial activities, and BI analysis, the use of management systems for corporate decision-making is less common. A corporate performance management (CPM) system is a tool that can assist businesses in dealing with such challenges. In todays' knowledge-based economy, it is of utmost importance for any firm to actively measure, monitor, and analyze their performance in order to gain a competitive advantage (Vukšić et al., 2013). In this scenario, CPM, also known as business performance management or enterprise performance management, plays a significant role in providing corporate managers a set of management strategies and technologies that enable businesses to stay ahead in the competitive environment (Frolick & Ariyachandra, 2006). Because corporate managers set the organization's strategy and facilitate its execution, their management practices determine organizational success (Simons, 2013). If we look into the case studies of former Fortune 500 companies, such as Kodak, Nortel, and Circuit City, they reveal that poor data collecting, processing, and analyzing, especially in the area of business strategy, are mainly responsible for their failure (Moore, 2010; Richards et al., 2014; Romero, 2013). Looking into the importance of data processing and analyzing, BI started gaining attention in the corporate world. According to Negash and Gray (2008), BI can aid in the improvement of operational procedures and also can be a very effective support to CPM.

BI is a broad group of programs that harvest and process information from a database, enable data visualization, and allow people to access subsets of data along multiple realms (Chen et al., 2012). The operational and technical infrastructure that gathers, retains, and analyzes the data generated by a company's activities is referred to as BI. Descriptive analytics, performance benchmarking, data mining, and process analysis are all covered under the umbrella of BI (Kautish, 2008; Kautish & Thapliyal, 2013). BI takes all of a company's data and organizes it into easy-to-understand reports, trends, and performance metrics that help managers in better decision-making (Bagale et al., 2021). In the world of chaotic data, BI brings clarity. BI is about executing a plan to obtain more insight into a company's data, be it data visualization or data-warehousing solutions. Although it may sound intimidating, BI is not just for the big businesses but for each and every size of a business, as its main purpose is to empower data and help companies to derive more benefits out of it.

2.2 History of BI

In the Cyclopedia of Commercial and Business Anecdotes, Richard Millar Devens coined the term *Business Intelligence* (BI) in 1865. He coined this term when he was describing how banker Sir Henry Furnese gained profit from the data that he had collected from the market and then acted upon it ahead of his competitors. More recently, in 1958, an IBM computer scientist named Hans Peter Luhn published an article detailing the possibilities of using technology to gather BI (Luhn, 1958). Only people with exceptionally specific talents could turn data into useable information in 1968. Data from numerous sources were often maintained in silos at the time, and analysis was presented in the form of an open-ended report. This was highlighted as a concern by Edgar Codd, who wrote a paper in 1970 that changed how people viewed databases. His idea of creating a "relational database model" acquired a lot of attraction and was adopted across the world.

The first database management system was called Decision Support Systems (DSS). The present version of BI, according to many historians, originated from the DSS database. As business people realized the significance of BI in the 1980s, the number of BI companies expanded. During this time, a variety of technologies were developed with the purpose of making data access and organization easier. Some of the technologies designed to work with DSS include data warehouses, Executive Information Systems (EIS), and OLAP (online analytical processing).

2.3 BI and Its Components

Today's organizations have more data at their disposal now than before. With feedback from customer surveys to production and distribution statistics, businesses generate, gather, and store massive volumes of data. BI is a set of strategies for putting this data to work in order to help firms become more efficient and profitable. Smart corporate executives may capture the power of raw data and use it to support organizational strategy that will help them stay ahead of the competition by employing these approaches and specific software analytics. The most significant components of BI are as follows.

2.3.1 Online Analytical Processing

OLAP is a system that allows users to examine information from various sources while providing numerous viewpoints or insights. A multidimensional data model is used in OLAP databases to facilitate complicated analysis and informal queries. OLAP's standard applications include sales and marketing reporting, financial and non-financial reporting, business process management, planning and budgeting, etc.

The diversity of options it offered to compile and organize information made OLAP highly popular. Its popularity as a SQL-based application faded as NoSQL gained traction. (At the moment, certain businesses, such as AtScale, Platfora, and Kyvos Insights, have overlaid OLAP on top of a NoSQL foundation.) Three operations are supported by OLAP and consolidation.

Slicing and Dicing: Slicing and dicing allow users to extract (slice) certain data from an OLAP cube and then view (dice) those slices from multiple viewpoints.

Drill-Down: Drilling down allows users to navigate and research information. For example, an auto-sales can be viewed by its gas usage, style, and color.

Consolidation: It is the process of bringing together data that may be stored and analyzed in a variety of ways. The automobile sales manager, for example, can add all the automobile sales from different branches in order to forecast the sales patterns.

2.3.2 Executive Information Systems

CEOs started to use the internet to investigate company data in the late 1970s. As a result, a software called EIS was developed to assist higher management in making choices. It is intended to give the most relevant and up-to-date data to expedite decision-making. When it comes to delivering information, the system mostly emphasizes visual displays and simple user interfaces.

2.3.3 Data Warehouses

With the popularity of Data Warehouses in the 1980s, companies started employing in-house Data Analysis solutions regularly. Since there were restrictions on computer systems, this was frequently done on weekends and after 5 p.m. on every weekdays. Before the popularity of data warehousing, a lot of repetitions were required to deliver meaningful information to multiple staff at the corporate level who were involved in making strategic decisions. Data warehousing reduced the time it took to obtain data by a substantial amount. Data warehouses aided in the development of Big Data applications. Instantly, a large amount of data in a range of formats could be retrieved from a single database, saving time and resources while also providing access to BI, which was previously not available. Data warehouses had enormous promise for providing data-driven insights that consequently help in improving profit base and early detection of frauds, and losses can also be reduced to a greater extent.

2.3.4 CPM or Advanced Analytics

This suite of tools enables executives to examine statistics for particular products or services. A fast-food business, for example, might study the sales of some products and make some necessary changes to its menu board options at the local, regional, and national levels. The information might also be used in anticipating the best possible market for the product to get success.

2.3.5 Real-Time BI

This particular aspect of BI is gaining popularity in a knowledge-based economy. As email, messaging systems, and even digital displays become more prevalent in real-time, businesses may adapt to these developments by utilizing software programs. A businessman may offer unique deals that reap the benefits of what's happening right now because everything is in real-time. For example, a voucher for an ice cream on a sunny day might be created by marketing experts using data. Customer interaction with a website may be tracked by corporate managers so that the sales-marketing experts may give special deals in real-time when the client is visiting the site.

2.3.6 Data Sources

There are several different types of data that are kept in this part of BI. Take the unprocessed data and utilize software programs to turn them into relevant sources of data that each department may use to positively influence a company. By employing this technique, BI experts may produce a huge number of workbooks, scatter plots, charts, and graphs that can be utilized for a wide variety of business objectives. Data, for example, may be utilized to build presentations that assist to create achievable team objectives. Organizations may make better-informed judgments if they consider the strategic element of data sources (Figure 2.1).

2.4 BI Tools

Numerous technologies are used to analyze the different elements of BI and turn them into practical problem-solving activities under the notion of BI. Our free-market economy's business insight has created specialized markets and start-ups, as well as consultancy companies and other commercial endeavors that have contributed to the development of various BI tools. It's possible to gain a deeper understanding of the firm by using these business-specific tools. Streaming analytics, prescriptive analytics, predictive analytics, descriptive analytics, and visual analytics are some of the most often used technologies today.

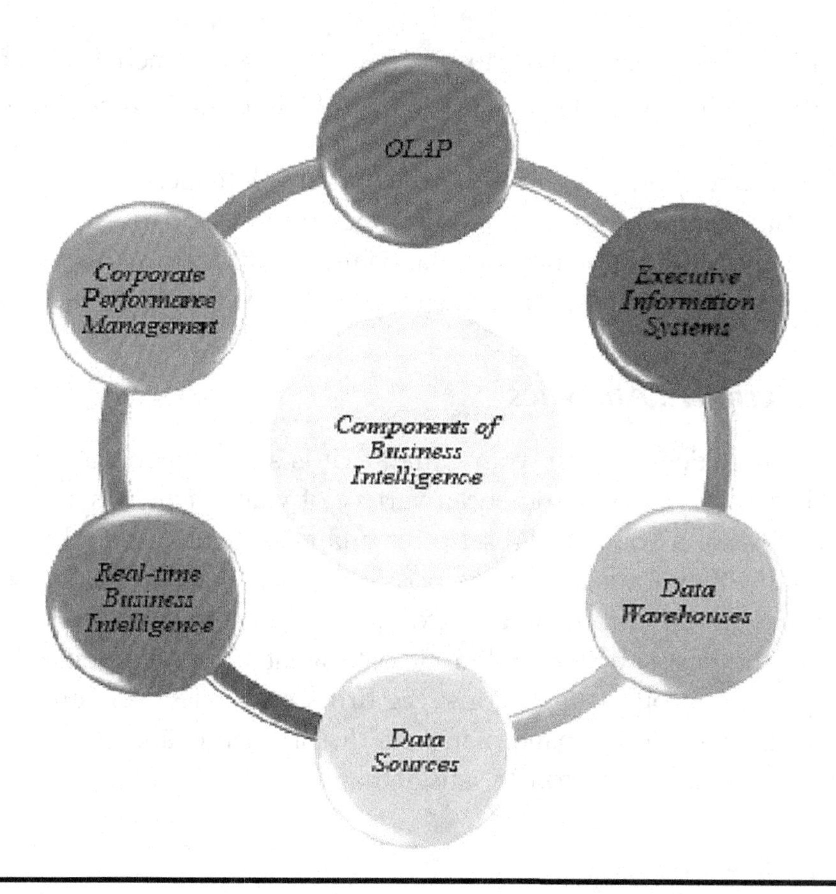

Figure 2.1 Business intelligence and its components.

2.4.1 Streaming Analytics

Data-based statistics is continually calculated, monitored, and managed in real-time as part of Streaming Analytics so that corporate managers can act upon it before their competitors in the market. This process entails being aware of what is happening in the market at any given time and taking action accordingly. They have enhanced the availability of valuable insights for decision-making. Cell phones, the Internet-of-things, market information, transactions, and devices, such as tablets and laptops, are all sources of data for Streaming Analytics. It links managers to other data sources, enabling applications to integrate and blend data into a flow of application, with relevant information, rapidly and effectively. Streaming Analytics is compatible with the following:

1. To minimize the harm that can be affected by social media breakdowns or cyberattacks or airline disasters or production faults or stock market crashes;
2. The real-time monitoring of regular company activities;
3. Using Big Data to uncover missing opportunities; and
4. There is also the possibility of establishing new business opportunities, investment opportunities, and technological developments.

2.4.2 Prescriptive Analytics

Prescriptive analytics is a very young area that is still a bit complicated to handle. These analytics recommend a variety of viable activities and direct individuals toward a solution. Prescriptive analytics is all about advising. Fundamentally, they forecast various possibilities and enable businesses to analyze a wide range of probable outcomes based on their activities. This analytics, in the best situation, will anticipate what would happen, why that would happen, and offer suggestions. Big firms have effectively employed prescriptive analytics to manage planning, distribution channels, and inventories, thereby ensuring customer satisfaction.

2.4.3 Predictive Analytics

Predictive analytics foresees the future. They utilize statistical data to provide businesses with valuable insights about impending developments, such as determining daily sales and purchase, and anticipating customer preferences. Typical corporate applications comprise projecting revenue growth at the end of each year, determining which goods buyers could buy at the same time, and projecting stock figures. Credit ratings are an example of this sort of analytics since financial firms use them to assess a customer's likelihood of making timely payments.

2.4.4 Descriptive Analytics

Descriptive analytics is mostly concerned with historical data and describes or summarizes data. They explain the past, helping us to see how earlier actions have influenced the present. This analytics may be used to describe different parts of a business and to explain how it functions. In the best scenario, this analytics presents a tale with a compelling subject while still providing valuable data (Figure 2.2).

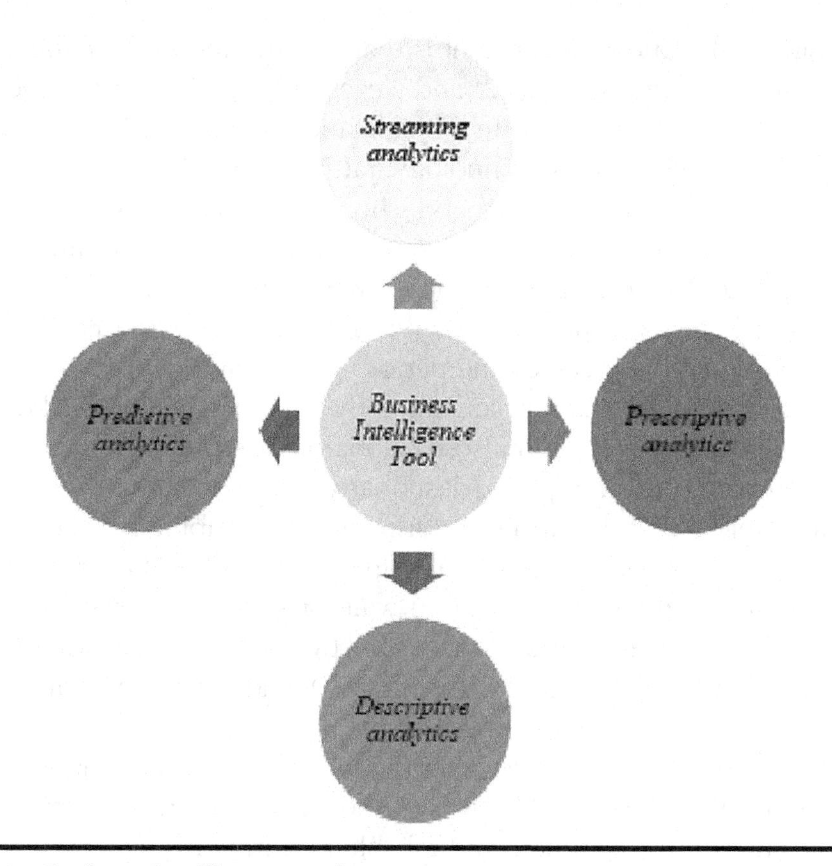

Figure 2.2 Business intelligence tools.

2.5 Role of BI in Organizational Growth

BI enables executives to make effective and relevant decisions based on facts. Organizations may better use data for competitive benefits by using solutions generated from BI components. When used correctly, data may help businesses make proactive decisions in response to market changes and other external variables. Although most organizations gather and retain vast volumes of raw data, only a small percentage of them really use that data to generate business insights and changes (Singh, Singh & Karki, 2021). BI procedures assist corporate managers in organizing their data so that it can be accessed and evaluated quickly. Decision-makers may then go deeper and swiftly obtain the information they want, helping to make possibly the best decisions. However, enhanced critical thinking is only one advantage of BI. In this episode, we'll look at the most positive implications of BI, as well as how firms are utilizing technology to reach their strategic goals. More than merely software, BI encompasses a broad range of activities.

In business, the one true constant is that it is continuously changing. According to Chen and Siau (2012) and Negash (2004), BI technologies are made to report on performance in a systematic manner. A BI system isn't made up of just one piece of technology; rather, it's made up of a variety of them. Past research, on the other hand, shows that OLAP is a key technology that enables decision-makers to view data from many angles (Ramakrishnan et al., 2012; Negash, 2004). For instance, any person can wish to look at sales of a certain product and then dig down to learn more about sales in a specific location or over a specific time period. This multidimensional examination of performance data is supplemented by technologies that make it easier to distribute online reports and also dashboards or scorecards (Chen & Siau, 2012). When compared to reports, dashboards or scorecards give decision-makers an overview of the firm's performance in a graphic format that enables variance analysis by managers. Since BI allows end users to explore performance data rather than having to wait for statistics from the technical division, it allows for quicker and more effective access to performance metrics. Negash (2004), Watson and Wixom (2007), and Sahay and Ranjan (2008) all claim that it can make the analysis of data easier and therefore enhance the management's capacity in deriving the meaning from the data supplied. As a result, BI likely has an influence on the CPM cycle in a variety of ways. It facilitates the supply of timely and reliable performance data, which has a direct influence on strategy and assessment. Additional data modification capability provided by BI might have a significant influence on analytics.

With the continued expansion of e-commerce, which has saturated every industry, BI has become more vital than ever. Anything a customer wants is only a click away, and for large corporations, this involves making wise judgments and understanding where and how to use their marketing budget (Singh & Gite, 2015). BI aids in the direction of these choices. There are a number of advantages of BI that include performance enhancement, accelerating sales and marketing, and helping in developing and maintaining a sustainable relationship with stakeholders by providing better customer satisfaction. In practice, BI adds value to an organization in the following ways.

2.5.1 Improved Speed of Analysis and More Understandable Dashboards

BI solutions are built to handle large amounts of data in the cloud or on your company's premises. BI systems collect data from many sources and store it in a database system, where it is analyzed using dashboards, drag-and-drop reports, and user queries. By using BI, Lenovo increased its efficiency in reporting by 95%. Their human resources department consolidated numerous monthly data into a single brief display. Again by using BI, PepsiCo also reduced 90% of their analysis time. BI dashboards have the advantage of making data analysis more simple and transparent and, thus, help people with no knowledge of coding to develop narratives with available data.

2.5.2 Reliable and Well-Managed Data

Data management and analysis are improved using BI systems. In conventional data analysis, data of different departments are segregated and users must access multiple databases to solve reporting issues. But, current BI solutions can now connect all the internal databases with external databases, such as user information, demographic information, and even past weather information, to create a single data warehouse. The very same information can be accessed by many departments at the same time. Tinuiti, a marketing agency, used BI technologies to consolidate over 100 databases, saving lots of valuable time of their clients.

2.5.3 Increases Customers' Satisfaction

Customer satisfaction and experience may be directly influenced by business information. An American multinational company, Verizon Communication, implemented BI tools across several divisions, leading to over 1,500 employee dashboards. The data for these dashboards came from processes, as well as text from customer assistance discussion forums. Verizon was able to use this information to discover ways to cut support calls by 43% and enhance service to customers.

2.5.4 Higher Level of Employee Satisfaction

IT personnel and analysts spend a little less time reacting to corporate user requirements. Divisions that previously had to rely on analysts or IT for information may now perform data analysis with less training. BI is built to be scalable, so it can provide data solutions to departments that need them as well as people who desire it. Tableau was expanded up to 1,000 worldwide users at the Brown–Forman Corporation, and it worked well with their current data structure. For non-technical people to look at data, BI software should provide a fluid and straightforward user experience.

2.5.5 Enhances Organizational Effectiveness

Leaders may use BI to access data and obtain a comprehensive picture of their operations, as well as measure outcomes against the rest of the company. Leaders may discover areas of potential by taking a comprehensive perspective of the organization. Pfizer utilizes BI tools to cooperate across departments and has built models to improve patient diagnosis and conduct clinical trials faster and more effectively. PEMCO, an insurance business, utilized Tableau to handle and close claims quickly. When companies spend less time analyzing data and preparing reports, they have more opportunities in developing new programs and products for their company.

2.5.6 Decisions Based on Data

Better company decisions may be made with reliable data and faster reporting capabilities. MillerCoors, a beer brewing company in the United States, created unique mobile dashboards for its sales staff, allowing them to view real-time data and sales predictions before meeting with prospective clients. They may confidently discuss the demands of clients or prospects since they know the data are accurate. Managers no longer are required to wait for days or weeks for reports or deal with the danger of obsolete information.

2.5.7 Competitive Advantage

Organizations can be more competitive if they understand the market and how they operate within it. Rosenblatt Securities, which is a research and investment banking boutique and agency-only institutional brokerage firm, evaluated information from a number of sources to determine the optimal moment to join and exit the market and to strategically position themselves in the market. Businesses may use BI to stay up with industry developments, track seasonal market shifts, and predict client demands. With the proper use of data and the conversion of data into actionable information, BI may help organizations gain a competitive advantage. BI systems make data available to approved clients and enable them to engage with competitive information from a single, secure, and centrally controlled database system. This enables businesses to make better strategic decisions by allowing them to obtain information quickly and efficiently (Figure 2.3).

In this knowledge-based economy, where businesses create, acquire, and store massive amounts of data on a daily basis, if information is not saved

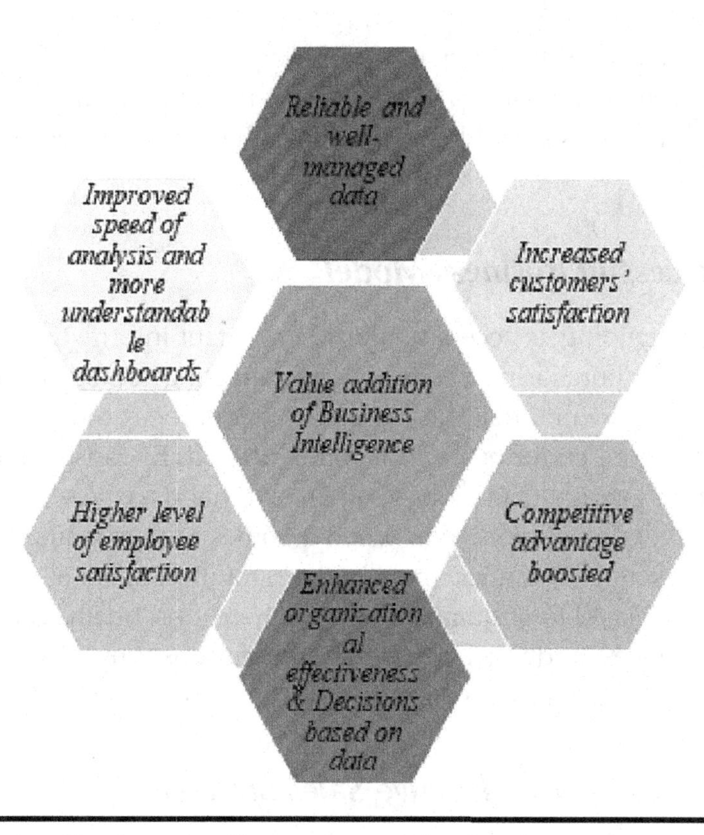

Figure 2.3 Role of business intelligence in organizational growth.

for future use, it becomes a threat to the business. Because the data analysis procedure is time-consuming, it is all too simple to misplace all of the data. However, BI optimizes and automates the whole data analysis process. This will help in making better business decisions. Further, there are more parameters in which BI plays a crucial role in enhancing business growth. These are as follows:

2.5.8 Enhance Market Intelligence

BI software may be used to gather market intelligence and data can be analyzed more effectively using BI tools, just as it can be used to improve competitive advantage. Identifying consumer needs, purchasing habits, tracking buyers' behavior, and anticipating market conditions are all part of BI. Organizations could be more sensitive to the market environment if all of these data are accessible from a single central location. This enables the top executives of the organization to organize their manufacturing unit.

They can also improve the efficiency of their product and decrease wastage. If the corporate manager or top executives of the organization can utilize BI wisely and smartly, this could help them in enhancing market intelligence in a number of ways.

2.5.9 A Successful Business Model

Once the organization gets access to all the important information, such as consumer needs, their purchasing behavior, and marketing strategy of their counterparts, it's indeed natural that the organization will focus more on making its business model more enduring and outcome-focused. BI assists in analyzing an organization's internal data along with its external data. Nowadays, every organization is very much particular when it comes to positioning their items, researching marketing dynamics, making the right investment at the right time for the activities related to sustainable development, and finally assessing their overall performances, and maybe these are the reasons why BI is attracting the interest of every organization's irrespective of their size and nature.

2.5.10 Composed and Effective Sales Strategies

It is constantly more vital for sales and marketing teams to comprehend the businesses with which they are attempting to collaborate. Before pitching a firm, every salesperson must study about numerous elements of it. Their yearly turnover, sales figures, merger data, rivals, trade policies, and outsourcing partners, among other things. The salesperson may readily obtain all of this information if the firm is publicly listed. If the salesperson doesn't have access to a computer, then they will have to rely on their own devices. BI will improve the coherence and effectiveness of your sales tactics. Before pitching any lead or organizing a business meeting, it is usually beneficial for the sales representative to know the business leads. BI solutions may obtain insight into the target business's turnover, financial plans, prospective development plans, sales data, rivals, and even more. This investigation may provide the salesperson with extra facts to evaluate and build a sales proposal. In the absence of BI, the sales managers will constantly struggle to anticipate which sales approach to apply in order to convert the business lead or close the transaction swiftly. Sales cannot be predicted; they must be achieved through investigation and effective execution. Sales experts utilize BI to make changes to their approach in creating sales pitches that are more detailed and personalized.

2.6 Integration of BI and Business Process Management

The notion of BI is not new and has been used by numerous firms and organizations. It has recently grown even more prevalent since it encompasses concepts, such as analytics, large data, and artificial intelligence, as vital components of digital processing, a crucial subject for corporate leaders of organizations and sectors of all kinds. BI-based decision-making is useful for firms with the main objective of boosting their organizational performance. However, research carried out by Audzeyeva and Hudson (2016) and Popovic (2012) reveal that this promise is realized only where BI information is conveniently leveraged to improve decision-making, creativity, and adaptability that ultimately leads to value maximization. It is therefore understood that BI is merely a facilitator—allowing better decision-making by the company. Consequently, BI has an indirect effect on organizational performance. BI is simply one among a number of organizational performance-enhancing measures.

Business process management is a well-known technique for increasing organizational performance through enhanced process performance. The integration of BI with business process management efforts appears logical, since both have the same objective. For instance, using BI to manage cross-functional business processes may improve the efficacy of BI assets. In reality, however, these ideas are often executed by disparate teams, resulting in inadequate collaboration and therefore misaligned efforts. Aligning a BI effort with a business process management initiative means that they both collaborate and contribute to the achievement of a company's objectives.

Although companies are increasingly recognizing that BI can contribute to enhanced business performance, Olszak (2016) articulated in their study that companies continue to undervalue the ability of BI to initiate effective decisions for enhancing business performance and processes. Dokhanchi and Nazemi (2015) argued that BI is often seen as a "technical notion" with minimal relation to business operations and that its strategic significance is largely ignored. According to Williams (2008), BI initiatives are often carried out by information technology departments and are most frequently carried out independently of corporate strategy and performance management projects. Suša Vugec et al. (2020) in their study have found that there is a lack of alignment between BI and business process management. According to the author, the lack of alignment is probably due to the fact that, on average, BI systems are developed as a technical effort, whereas process management is recognized and executed as an enterprise management initiative. Thus, the divergence between theory and practice in this area is simply a reflection of the divide between information technology governance and the rest of the company (Peppard, 2001). Thus, a gap may be bridged by giving information technology a greater operational and organizational role and by recognizing that developing a BI system is mainly a business-focused initiative, not an information technology-focused one (Jude et al., 2021). When business process management is kept at the process modeling level, i.e., at the operational level, project teams lack the authority to bring about alignment.

2.7 Conclusion

Nowadays, we are all part of the corporate environment in which companies gather and retain massive amounts of data from numerous sources. If the data are not used correctly, it becomes a burden and an expense for the company. It is necessary to assess each and every data point carefully in order to create company plans, but this is a time-consuming and costly exercise. Using BI, the operation may not only be improved but also streamlined. BI delivers data that are created for the business. Companies that use BI can quickly diagnose facts from vast amounts of unstructured data. With immediate access to competitor's company data, a corporate manager can analyze internal information and make more effective decisions. BI specialists guarantee that the firm receives real-time enhanced business insights so that it may efficiently use the information for a better outcome. A powerful BI system may save a significant amount of time spent on data gathering, input, analysis, control, and usage. These tasks require a significant amount of human time and effort. An automated BI solution can collect, analyze, organize, and use data more quickly and effectively. Reports may be created fast since all of the data are already in their proper position behind the scenes. It saves a significant amount of time and human effort.

References

Audzeyeva, A., and Hudson, R. (2016). How to get the most from a business intelligence application during the post implementation phase? Deep structure transformation at a U.K. retail bank. *European Journal of Information Systems*, Vol. 25 No. 1, pp. 29–46.

Bagale, G.S., Vandadi, V.R., Singh, D. et al. (2021). Small and medium-sized enterprises' contribution in digital technology. *Annals of Operations Research*. Doi: 10.1007/s10479-021-04235-5.

Chen, X., and Siau, K.L. (2012). Effect of business intelligence and IT infrastructure flexibility on organisational agility. *Thirty Third International Conference on Information Systems*, Orlando.

Dokhanchi, A., and Nazemi, E. (2015). BISC: A framework for aligning business intelligence with corporate strategies based on enterprise architecture framework. *International Journal of Enterprise Information Systems*, Vol. 11 No. 2, pp. 90–106.

Frolick, M., and Ariyachandra, T. (2006). Business performance management: One truth. *Information Systems Management*, Vol. 23 No. 1, pp. 41–48.

Jude, A.B., Singh, D., Islam, S., et al. (2021). An artificial intelligence based predictive approach for smart waste management. *Wireless Personal Communications*. Doi: 10.1007/s11277-021-08803-7.

Kautish, S. (2008). Online banking: A paradigm shift. *E-Business, ICFAI Publication, Hyderabad*, Vol. 9 No. 10, pp. 54–59.

Kautish, S., and Thapliyal, M.P. (2013). Design of new architecture for model management systems using knowledge sharing concept. *International Journal of Computer Applications*, Vol. 62 No. 11, pp. 27–30.

Luhn, H.P. (1958). A business intelligence system, *IBM Journal*, Vol. 2 No. 4, pp. 314–319.

Moore, S. (2010) *Profile of Kodak: From Film to Digital Photography*. Ann Arbor: University of Michigan, William Davidson Institute, 15 p.

Negash, S. (2004). Business intelligence. *Communications of the Association for Information Systems*, Vol. 13 No. 1, pp. 177–195. Doi: 10.17705/1CAIS.01315.

Negash, S., and Gray, P. (2008). Business intelligence. In: Burstein, F., and Holsapple, C., editors. *Handbook on Decision Support Systems 2: Variations*. Berlin: Springer Berlin Germany, pp. 175–193.

Olszak, C.M. (2016). Toward better understanding and use of business intelligence in organizations. *Information Systems Management*, Vol. 33 No. 2, pp. 105–123.

Peppard, J. (2001). Bridging the gap between the IS organization and the rest of the business: Plotting a route. *Information Systems Journal*, Vol. 11 No. 3, pp. 249–270.

Popovic, A., Hackney, R., Coelho, P.S., and Jaklic, J. (2012). Towards business intelligence systems success: Effects of maturity and culture on analytical decision making. *Decision Support Systems*, Vol. 54 No. 1, pp. 729–739.

Ramakrishnan, T., Jones, M.C., and Sidorova, A. (2012). Factors influencing business intelligence data collection strategies: An empirical investigation. *Decision Support Systems*, Vol. 52, pp. 486–496.

Richards, G., et al. (2014). A process view of organizational failure: The case of Nortel. *Academy of Management Proceedings*, Vol. 2014 No. 1, pp. 223–235.

Romero, J. (2013). *The Rise and Fall of Circuit City*. Economic History, Econ Focus-Richmond, Vol. 17, No. 3, pp. 31-33.

Sahay, B.S., and Ranjan, J. (2008). Real time business intelligence in supply chain analytics. *Information Management and Computer Security*, Vol. 16 No. 1, pp. 28–48.

Simons, R. (2013). *Levers of Control: How Managers Use Innovative Control Systems to Drive Strategic Renewal.* Boston, MA: Harvard Business Press.

Singh, A., and Gite, P. (2015). Corporate governance disclosure practices: A comparative study of selected public and private life insurance companies in India. *Apeejay - Journal of Management Sciences and Technology* Vol. 2 No. 2.

Singh, D., Singh, A., and Karki, S. (2021). Knowledge management and Web 3.0: Introduction to future and challenges. In *Knowledge Management and Web 3.0.* De Gruyter, Cambridge University Press. Doi:10.1515/9783110722789-001Agents.

Suša Vugec, D., Bosilj Vukšić, V., Pejić Bach, M., Jaklič, J., and Indihar Štemberger, M. (2020) Business intelligence and organizational performance: The role of alignment with business process management. *Business Process Management Journal,* Vol. 26 No. 6, pp. 1709–1730. Doi: 10.1108/BPMJ-08-2019-0342.

Vukšić, V.B., Bach, M.P., and Popovič, A. (2013). Supporting performance management with business process management and business intelligence: A case analysis of integration and orchestration. *International Journal of Information Management,* Vol. 33 No. 4, pp. 613–619.

Watson, H.J., and Wixom, B.H. (2007). The current state of business intelligence. *IEEE Computer,* Vol. 40 No. 9, pp. 96–99.

Williams, S. (2008). Power combination: Business intelligence and the balanced scorecard. *Strategic Finance,* Vol. 89 No. 11, pp. 27–35.

Chapter 3

Transforming HR through BI and Information Technology

Deepak Bangwal, Prakash Chandra Bahuguna, Rupesh Kumar, and Akhil Damodaran
University of Petroleum and Energy Studies

Jason Walker
University Canada West

Contents

DOI: 10.4324/9781003184928-3

3.1 Introduction

During rapidly changing atmosphere, the requirement of accurate, relevant and prompt information is not only essential for competitive advantage but also a key for sustaining the business. In response to the growing significance of data intelligence for executives and their business environments, today's organizations are making a good number of investments in business intelligence (BI) systems. The primary purpose of BI is to describe an organization's data assets for evolving a correct understanding of business dynamics and building informed conclusions on information assembled from numerous sources (Aruldoss et al., 2014; Li et al., 2008). Every organization has a different job context and scope, and they all face problems while using existing data, as existing data are always a valuable resource to dig out relevant knowledge and help managers to take important decisions in different business scopes. Nowadays, storing the data and analyzing the data in an effective manner is one of the most critical issues. Globalization, workforce diversity, environmental concerns and technological advancements have transformed the business environment. Robotics, artificial intelligence, machine learning, big data, augmented and virtual reality, etc. have become integral part of business environment. These environmental forces have created a kind of stormy situation for business organizations (Kautish, 2008; Kautish & Thapliyal, 2013). They are under tremendous pressure to be competitive and survive in the market.

On diving deeply, we see, actually, organizations do not compete for markets; they compete for human resource (Bagale et al., 2021), because in the time of fourth industrial revolution, human resource is the only resource that can be the source of sustainable competitive advantage. How well the organization responds to the challenges originating from multiple stakeholders, different constituents of business environment and in particular from unprecedented technological advancements, depend on the availability of agile, techno perceptive human resource who are ready to learn, unlearn and change quickly. In the era of digitization, automation, modularization and intelligentization (Lu et al., 2019), the human resource function needs to relook at its existing traditional HR architecture, as the current age business strategies require different sets of employee attitudes, behaviours and capabilities as the workplace has changed drastically. It seems that every organization, irrespective of its primary business, has become a technology firm. As a result, today's HR functions have changed from traditional service-oriented functions to technology-driven practices. The strategic contribution of HR is possible only when HR function embraces the developments, such as BI, happening at a macro level (Singh, Singh, & Karki, 2021). As we are aware that BI helps improve strategic and operational decisions in a more scientific way and with clear evidences, HR cannot think of not adopting and remaining aloof. It needs to develop its own intelligence system; HR can enhance evidence-based decision-making and develop into a true business partner. The basic objective of human resource intelligence system is to collect and analyze data for predictive and prescriptive purposes. With such a system, HR professionals will be able to assist management in the areas of people management and help them formulate strategic plans accordingly. A well-established human intelligence system provides senior managers with information and insights.

3.2 Understanding BI

BI helps business decisions with data and analysis cutting across all functions (e.g., marketing, production, supply chain, human resource). The key processes involved in BI are data collection, preparation, analysis, testing, implementation and optimization. BI provides information about what is going on in the business. Information is accessible in a timely and flexible basis with regard to products, markets, people, competitors, etc. When used effectively, BI helps organizations to improve revenue, margin and profitability. The tangible advantages of BI for business organizations include the following:

i. Help create value
ii. Better control
iii. Improve performance
iv. Improve operational efficiency
v. Improve process
vi. Improve customer service
vii. Identify and capitalize business opportunities

BI can be conceptualized and implemented in various ways. For example, Petrini and Pozzebon (2008) developed two basic approaches. First, at the philosophical level, is the managerial approach, grounded on the assumption of superiority of management decision-making, and the second one at the level of implementation. This second approach is technical in nature, wherein BI is viewed as an instrument that supports and supplements managerial decision-making. BI concepts and approaches can create a comprehensive and relevant information, which can help organizations, managers and supervisors to generate pertinent reports, which can further help in decision-making, whereas the traditional information systems can't generate such reports with accurate, comprehensive and timely information. Because, fetching all of the relevant data and information together on one page, and which is easy to interpret and to understand by the team leaders, supervisors, subordinates and other supporting staff members, is a very effective tool in today's competitive atmosphere. Through BI, managers are able to utilize dashboards effectively; they fetch accurate and relevant information instead of fetching irrelevant data and information. In today's era, organizations need to understand the worth of getting on-time information, by using this contemporary and valuable business tool. Drawing from the literature, we can conclude that prominently there are three important approaches to use BI.

i. **Managerial Approach**: Managerial approach concentrates on refining management decision-making.
ii. **Technical Approach**: Technical approach concentrates on tools, which assist the procedures linked with intelligence in management approach.
iii. **Enabling Approach**: Enabling approach concentrates on other upgraded competence in maintenance and managing of information.

3.3 Forces Influencing BI

As mentioned in the introductory section, there are a host of factors that have necessitated BI. Broadly, the factors that drive BI initiatives can be categorized into following categories:

i. Globalization
ii. Demographic changes
iii. Technological advancements
iv. Increased awareness and expectations of multiple stakeholders

There are enough evidences available in the literature and the managerial anecdotes that highlight how forces emanating from shrinking of global business boundaries, demographic changes (current composition of workforce comprise four generations: generation z, millennials, generation x), technological advancement (such as artificial intelligence, machine learning, robotics) and growing awareness about societal, environmental, economic performance and corporate frauds have put tremendous pressure on the organizations to be agile, resilient and socially responsible.

3.4 Implications for Human Resource Architecture

Every organization, irrespective of its size, product or nature has a HR architecture to support its business strategies and achieve its strategic goals. HR architecture is an organizational framework composed of a HR system to ensure that employees display desired behaviour patterns and possess required competencies in accordance with its business strategy or set of strategies (Stewart & Brown, 2019; Irfan et al., 2018; Purcell & Kinnie, 2007). Drawing from the review of literature (e.g., Boon & Lepak, 2019; Stewart & Brown, 2019), HR system can be defined as a combination of people (HR professionals), HR strategies, HR policies and HR practices. It is a well-known fact now that organizations need dynamic business strategies that fit the business environment, the stages of life cycle (e.g., start up, growth and maturity) and the context, which in turn requires HR to have a supportive system in place to help workforce exhibit different appropriate behaviours and competencies.

According to Vrchota et al. (2020), some of the most important skills and abilities that industry 4.0 has necessitated are "IT skills, ability to process and analyze data, ability to use latest devices, time management, excellent communication skills, adaptability to change and social skills". Competencies have also been classified as professional (e.g., problem solving), personal (e.g., systematic thinking) and social (e.g., adaptability). This has a clear mandate for HR to have strategies and practices that prepare the work force for these skills and abilities. Sima et al. (2020) posit that a unique combination of human intelligence; information technologies (ITs) and innovations are the crucial requirements of an Industry 4.0 era. This has a clear indication for new HR system that is designed to enhance the new age abilities and behaviours. The new age HR architecture needs to move from descriptive to prescriptive analytics.

3.4.1 Implications for Human Resource Practices

The trends have clear implications for HR practices as well. Bailey (1993) theorized that an employee's discretionary efforts are contingent upon skills, motivation and opportunity to perform, which later came to be known as AMO framework (Applebaum et al., 2000). AMO framework classifies HR practices into ability, motivation and opportunity enhancing practices. Although there is no consensus among researchers about the bundles of the practices that belong to these three categories, however certain practices, such as selective hiring, training and development (ability-enhancing), merit pay, incentives, recognition (motivation enhancing) and participative management, self-directed work teams (opportunity enhancing practices), are commonly cited HR practices. As evident from the ongoing trend, the nature of people management has already changed, for example, artificial intelligence-assisted workforce management, AI assisted hiring and succession planning, compensation management, performance management, etc.

3.4.2 Implications for HR Professionals

As discussed, in the previous sections, today's business environment is immensely different from the business environment of yesteryear, which has necessitated the changes in the mindsets, business models, strategies and sources of creating competitive advantage. Different periods require different orientations (Kautish, Singh, Polkowski, Mayura, & Jeyanthi, 2021). For example, during the subsequent years of industrial revolution, automation sought to replace manual work, where mass production became the mantra for survival. Later, with growing consumerism and focus on quality products, the production-based advantage shifted to customer orientation, including customer satisfaction, timely delivery and after sales. As a result, marketing started getting the attention of business enterprises. Later towards the twentieth century, when the world witnessed the crumbling of national boundaries, breakthrough in information and communication technology, new business models started coming up. Consequently organizations started understanding the role of human resource (skills, abilities, attitudes, behaviours, risk-taking abilities and adaptability) in creating sustainable competitive advantage that not only helps organizations to achieve its economic goals but also enables them achieve goals of survival and growth.

As companies are increasing their reliance on BI, all traditional business functions are becoming data-driven. However, the most hit function is human resource management, putting additional stress on human resource functionaries. Human resource professionals being an important component of any HR architecture need to reorient themselves and possess competencies that are relevant and appropriate to fast changing business landscapes, where speed, accuracy and alertness are the buzz words. HR functionaries need to possess a wide variety of skills for effective functioning. As strategic partners, they need business, strategic, analytical and statistical and problem-solving skills. For effective intelligence, they need to have a good understanding of research methodology. Some of the important BI skills that HR professionals need to possess are as follows:

3.4.2.1 Strategic Skills

According to CIPD, the best HR function is one that understands the challenges faced by the organization and provide data-driven insights in formulating and implementing business strategies. The strategic skills of HR managers help their organizations to integrate human resource planning to strategic planning.

3.4.2.2 Business Skills

Business skills refer to understanding the business models, value chain and understanding financial statements as these skills help them interpret business models, strategies and financial positions and formulate appropriate HR strategies.

3.4.2.3 Problem-Solving Skills

Closely related to business and strategic skills are the problem-solving skills. HR professionals would be able to provide intelligence only when they understand the problem, are able to ascertain the possible causes and provide a requisite solution. Problem solving and analytical skills improve evidence-based decision-making.

3.4.2.4 Research Skills

For effective HR intelligence, HR professionals are required to acquire and sharpen their research skills. In today's business environment, research skills are of extreme importance. Research skills refer to understanding of the research methodology so that they can recommend research-based solutions to the organizational problems.

3.4.2.5 Statistical Skills

Knowledge of statistics is the basic need for HR intelligence, as these skills help in the analysis of data and present information in a systematic manner. Many of the HR functions, such as performance management, hiring and compensation management, need application of statistical skills.

In addition to the BI skills, the other essential elements for effective delivery of HR services, employee and organizational outcomes are as follows:

3.4.2.6 HR Functionaries' Transformational Change

Besides the core BI skills, skills related to change management, leadership and negotiation form the largest skill component needed to understand and deal with today's business environment.

3.4.2.7 Organization of HR

The current or traditional organizational structure of HR departments need reorganization as the conventional structures may fail to respond to new age strategic challenges. Designing an intelligent human resource system is essential to deliver quick, accurate and complete solution and making non-programmed decisions.

3.4.2.8 Integration of Plans

HR function will never be able to provide competitive intelligence to top management unless the two different types of plans are interwoven. HR planning has to be integrated with strategic plans of the organization. Simply, this means the strategic integration between two levels of functioning. This integration not only helps the HR professionals to implement the strategic intentions of the top management but also helps them to act as information providers that enable the top management to formulate evidence-based decisions. Many authors (e.g., Tyson, 1985) argue in favour of this kind of vertical integration where HR assumes the role of a strategic partner by emerging as an intelligence provider.

3.4.2.9 HR and Line Partnership

In addition to change in the HR professionals' mindset, reorganization of HR structure and integrating HR and strategic planning; another important constituent of an effective HR system is line management. Ultimately, the intended HR policies need commitment and support of line managers and, on the other hand, HR professionals would be able to provide needed assistance to line managers only when these two work together in tandem. Without each other's inputs, the entire organization fails to create the desired organizational culture and achieve strategic objectives.

3.4.2.10 Benchmarking

Benchmarking is an essential requirement for BI. Unless the HR system benchmarks its policies, practices and strategies against industry standards, it cannot discharge its role in BI. The HR systems need robust measurement tools, techniques, measures and provisions in place that measures the impact of HR initiatives on employees and organizational outcomes. Benchmarking and measurement enable organizations to make intelligent and evidence-based decisions.

3.5 HR and IT

As mentioned, there are a variety of factors, such as globalization, diverse workforce and advancements in manufacturing technology, that have transformed the world around us, but high on the list are unprecedented advancements in IT as it has multifaceted implications. We know how the Internet, ERP, analytics and cloud computing have transformed the nature of business. IT could make work from home possible. Free flow of goods, people and capital could become a reality only because of IT.

Human resource being a core management function ought to catch the wave and capitalize technology and analytics to develop a robust intelligence system. HR managers need to develop their capabilities and competencies to implement the new age software, databases, technology and decision-making tools to make fast and strategy-enabling decisions.

Although electronic HRM is not a new term, organizations have been using eHRM for administrative purposes, yet it has not acquired a much-awaited place in literature. The simple reason for this is due to the lack of its strategic contribution (Lazazzara & Galanaki 2018). How well organizations and HR departments integrate IT, moderates a number of organizational parameters (such as efficiency, effectiveness, innovation) and determines profit, survival and growth. As suggested by Ulrich (1998) and Tyson (1985), congruence between HR and strategy is important; in the same way, adoption of IT hence integrating it with conventional human resource tasks and activities helps HR to become strategic in nature, which is not just a support function but truly can become an intelligent function.

IT–HR congruence refers to embedding technology into conventional HR functions, which support each other's objectives and thus help organizations to contain costs and make data-driven decisions.

A good HR information system has multiple advantages at the operational and strategic levels. Some of the advantages of an efficient HR information system that are frequently cited in the literature are effective human resource planning, employee turnover modelling, intelligent hiring, effective training need analysis and strategic compensation planning. A good human resource information system provides data warehousing, data analytics and insights for solving administrative problems and making more intelligent business decisions.

3.5.1 Factors Affecting HR and IT Fit

Drawing upon the literature, we can categorize the factors influencing IT and HR fit, broadly into four categories: organizational, technological, environmental and individual (Bhattacharya & Wright, 2005; Lado & Wilson, 1994; Lado, Boyd & Wright, 1992; Wei & Lau, 2005). To understand the factors, the existing systems theories are of great help. The open system theory clearly provides an insight about the factors that have bearing on the systems. Various actors outside the HR system exert pressure on the system and provide inputs for creating the desired alignment between these two distinct areas. These factors include environmental and situational factors. Environmental factors include changing business contexts, technological advancements, government regulations, demographic changes happening at the international level, etc., outside the organizations. Similarly, another set of factors that influence the IT and HR fit belong to the environment, which are outside the system but within the same organization. These factors relate to business strategy, organizational structure, organizational culture, attitude towards technology adoption and management philosophy. At the lower level, there are factors related to the HR sub-system itself. The philosophy, the existing processes and the position of the sub-system within the larger system of organization do influence the congruence. Additionally, the people, HR professional, top management team, line managers and employees do influence the IT and HR congruence; their values, orientations, personality, leadership style, attitudes, needs, motivation and goals. All these factors are interlinked, interacting and interdependent and therefore have a synergistic effect on the possibility, level and type of congruence or alignment between HR and IT. Therefore, a complete assessment of all the factors is required. Dewett and Jones (2001) framework on the role of IT in the organization provides clear boundaries of IT and HR alignment. The framework suggests IT as the moderator of relationship between two main components of organizations – i.e., (1) organizational characteristics – outlining structure, size, learning, culture and interorganizational characteristics as key elements – and (2) organizational outcomes – where organizational efficiency and innovation are created. The addition of environmental and situational factors would make the framework complete. Consideration of situational factors is very important as they keep on varying with the context and the stage of business on the life cycle curve of the organization.

3.6 Conclusion

Drawing from the literature, we can conclude that BI is the need of the hour. It is a unique and scientific way of utilizing information and examining organizational processes and systems, which further helps to support decision-making. BI can help top management, business analysts, managers and subordinates to predict the behaviour of competitors, traders, clients and environments to stay alive and survive in a rapidly changing environment. BI provide support to data-driven and evidence-based decision-making. The decisions taken on BI prove to be timely, accurate, precise and complete. All levels of management get benefits from intelligence-enabled decisions.

Hence, in this study, we systematically review the recent research and academic papers in BI scholarship to classify and prioritize the concepts and perspectives of BI. Further, the outcome of this study intends to facilitate leaders, supervisors and managers to better recognize the importance, constraints and develop organization-specific intelligent systems to improve organizational opportunities to optimize profits and increase the chances of survival and growth.

References

Applebaum, E., Bailey, T., Berg, P., & Kalleberg, A. (2000). *Manufacturing Advantage: Why High-Performance Work Systems Pay Off.* Ithaca, NY: Cornell University Press.

Aruldoss, M., Lakshmi Travis, M., & Prasanna Venkatesan, V. (2014). A survey on recent research in business intelligence. *Journal of Enterprise Information Management*, 27(6), 831–866.

Bagale, G.S., Vandadi, V.R., Singh, D. et al. (2021). Small and medium-sized enterprises' contribution in digital technology. *Annals of Operations Research*, Doi: 10.1007/s10479-021-04235-5.

Bailey, T. (1993). *Discretionary Effort and the Organization of Work: Employee Participation and Work Reform Since Hawthorns*. Working Paper, Columbia University.

Bhattacharya, M., & Wright, P.M. (2005). Managing human assets in an uncertain world: Applying real options theory to HRM. *The International Journal of Human Resource Management*, 16(6), 929–948.

Boon, C., & Lepak, D.P. (2019). A systematic review of human resource management systems. *Journal of Management*, 45(6), 2498–2537.

Dewett, T., & Jones, G.R. (2001). The role of information technology in the organization: A review, model, and assessment. *Journal of Management*, 27, 313–346.

Kautish, S. (2008). Online banking: A paradigm shift. *E-Business, ICFAI Publication, Hyderabad*, 9(10), 54–59.

Kautish, S., Singh, D., Polkowski, Z., Mayura, A., & Jeyanthi, M. (2021). *Knowledge Management and Web 3.0: Next Generation Business Models*. Berlin: De Gruyter.

Kautish, S., & Thapliyal, M.P. (2013). Design of new architecture for model management systems using knowledge sharing concept. *International Journal of Computer Applications*, 62(11), 11–15.

Lado, A.A., Boyd, N.G., & Wright, P. (1992). A competency-based model of sustainable competitive advantage: Toward a conceptual integration. *Journal of Management*, 18(1), 77–91.

Lado, A.A., & Wilson, M.C. (1994). Human resource systems and sustained competitive advantage: A competency-based perspective. *Academy of Management Review*, 19(4), 699–727.

Lazazzara, A., & Galanaki, E. (2018). E-HRM adoption and usage: A cross-national analysis of enabling factors. In: Rossignoli C., Virili F., Za S. (Eds) *Digital Technology and Organizational Change. Lecture Notes in Information Systems and Organisation*, 23. Springer, Cham. Doi: 10.1007/978-3-319-62051-0_11.

Li, S.-T., Shue, L.-Y., & Lee, S.-F. (2008). Business intelligence approach to supporting strategy-making of ISP service management. *Expert Systems with Applications*, 35(3), 739–754.

Lu, H., Guo, L., Azimi, M., & Huang, K. (2019). Oil and gas 4.0 era: A systematic review and outlook. *Computers in Industry*, 11, 69–90.

Petrini, M., & Pozzebon, M. (2008). What role is "Business Intelligence" playing in developing countries? A picture of Brazilian companies. In: Rahman, H. (Eds.), *Data Mining Applications for Empowering Knowledge Societies*, IGI Global, 237–257 (Chapter XIII).

Purcell, J., & Kinnie, N. (2007). Human resource management and business performance. In P. Boxall, J. Purcell, & P. Wright (Eds). *The Oxford Handbook of Human Resource Management.* Oxford: Oxford University Press.

Sima, V., Gheorghe, I. G., Subić, J., & Nancu, D. (2020). Influences of the industry 4.0 revolution on the human capital development and consumer behavior: A systematic review. *Sustainability,* 12(10), 4035.

Singh, D., Singh, A., & Karki, S. (2021). Knowledge management and Web 3.0: Introduction to future and challenges. In *Knowledge Management and Web 3.0: Next Generation Business Models,* De Gruyter, 1–14. Doi: 10.1515/9783110722789-001.

Stewart, G. L., & Brown, K. G. (2019). *Human Resource Management.* John Wiley & Sons.

Tyson, S. (1995). *Human Resource Strategy: Towards a General Theory of Human Resource Management.* London: Pitman.

Vrchota, J., Marikova, M., Rechor, P., Rolinek, L., & Tousek, R. (2020). Human resources readiness for industry 4.0. *Journal of Open Innovation: Technology, Market, and Complexity,* 6, 3.

Wei, L., & Lau, LC.M. (2005). Market orientation, HRM importance and HRM competency: Determinants of SHRM in Chinese firms. *International Journal of Human Resource Management,* 16(10), 1901–1918.

Chapter 4

The Role of Business Intelligence in Organizational Sustainability in the Era of IR 4.0

Lakshmi C. Radhakrishnan
Institute of Management & Technology (IMT Business School)

S.B. Goyal
City University

Pradeep Bedi
Galgotias University

Vijay Nimbalkar
International Institute of Management Studies

Contents

DOI: 10.4324/9781003184928-4

4.1 Introduction

Business intelligence (BI) and data management technologies have been supporting global organizations since almost a decade to reliably collect, disseminate, interpret and store a huge volume of information, which has made information access and security effective. This effectiveness has urged organizations to adopt innovative technological interventions to manage data and utilize it in the most apt situations, thus also creating job opportunities for knowledge workers with high analytical skills.

The term BI was coined in 1958 by H.P. Luhn, who combined the terms business and intelligence. Although intelligence is knowledge, skill and competence to work towards a desired set of actions, when the same intelligence is applied to business, science, law, commerce and business information, the new field of study is termed as Business Intelligence (Luhn, 1958). Morris (2018) defined BI as a framework that will transform business-relevant data into knowledge and wisdom.

As a Vice President of Gartner, through his research, Howard Dresner in 1989 defined BI as a combined term that incorporates the concepts and techniques that utilize data-driven techniques to take improved decisions (Dresner et al., 2002).

The term BI can be defined in many ways, but the essence of the BI systems leads to the processes involving collection, sorting and retention of volatile information to support easy retrieval and support high-quality decisions. BI is a process involving raw data collection from multiple yet reliable sources for future predictions of events and informed decision-making, and during this process, the collected data are analyzed for patterns through a data-mining process.

BI is growing to be an answer to the complexities faced by contemporary organizations. BI systems are expected to support smart exploration, data integration, accumulation and analysis of data from allied sources of information that are reliable and up-to-date to resolve complex issues in different areas of management (Singh, Singh & Karki, 2021). Research reveals that BI systems lead to increased levels of information clarity, improved transparency and knowledge management that enable organizations to (Kalakota & Robinson, 1999; Liautaud & Hammond, 2002; Moss & Alert, 2003):

- collect information on organizational revenue streams and profitability;
- expenditure and assessment of return on investment;
- monitor and manage corporate stakeholders and business environments; and
- determine possible business incongruities and frauds.

With relevance to managerial perspective, BI is a combined effort from individuals/teams handling multiple processes. The roles (Grossmann, 2015) include

- **Process Owner** who sets the goals and the rules to attain the goals.
- **Process Subjects** who identify the process flow and instances to be followed.
- **Process Actors** who are executors of the planned events.

This chapter extensively adopts the term BI to introduce, discuss and explain the use of BI as a technology intervention that aids easy information access, transformation, storage and retrieval of extensive and diverse data (Chen et al., 2010, 2012).

4.2 Overview of BI: Evolution and Growth

BI is often referred to as innovative techniques, data processing systems, technology methodologies and tactful applications used to assess information and analyze data to help organizations make improved decisions (Chen et al., 2012, p. 1166), and BI has been comprehensively described as the ideas and approaches used to make improved decisions in business contexts using information and fact analysis systems, including fundamental tools, techniques, constructs, applications, databases and techniques (Chen et al., 2010, p. 201).

The initial terminology of BI was coined by Hans Peter Luhn in 1958.

Although a very prominent member in data security and trend prediction, the origins of BI have a humble start with data interpretation using COBOL to generate application reports. Although at the time, application reports served data interpretation effectively, the validity of data in the dynamic environment was questionable and the number of printed sheets generated by COBOL and application reports took a great amount of time to run. The BI is an evolved technology from Data Management Systems used in the past.

But as constraints and challenges came in to improve validity and ease of data access, experts resorted to online transaction and reporting. Although online transactions allowed organizations to interpret data in real time and up-to-date, the online transactions were not good to represent large amounts of data as the transactions kept changing second by second and the access of data was limited to specific commands and applications fed into the system as commands, which meant only one particular data was available at a time, which meant data restriction and more time for more data. Further, if two users requested the same data at two time intervals, the data would change as it is in real time.

Thus, it meant, for large organizations, where specific, time-based and integrated data were required, the multiple overlapping data would be challenging with no support to decision-making. This evidently led to the creation of data warehouses. The data warehouse led to a new style of reporting, analysis and data access that came into existence and with this, the new paradigm of BI was born.

4.2.1 BI, Business Analytics and Artificial Intelligence: All About Data and Beyond

BI is often confused with business analytics (BA) and artificial intelligence. Although all the above systems fall under the new trends of data management and recall in the new digital era, it is worth making an effort to clearly differentiate these terms from each other.

Business Intelligence is a descriptive form of data access and trend analysis that tells us what is happening at present and what has occurred in the past that has led to the present state.

Business Analytics is in turn predictive in nature and prescriptive to inform us based on the trends generated of what could be the future trends and outcomes of specific actions being implemented. Thus, analytics informs us what needs to be done at present to have a forecasted outcome.

Artificial Intelligence involves any and all technologies that use computers to mimic attributes of human intelligence, including sensory perceptions, problem solving and decision-making based on learning. AI in business effectively supports automation of business data for the visualization of inventory and/or sales data by combining large datasets, pattern analysis in real time. AI is also technology integrated into BI for analyzing customer needs and buying trends and satisfaction.

4.2.2 Application of BI: The Paradigm Shift and Process Flow

The terms business analytics, business intelligence and big data management are often interchangeably used to define the set process of information filtration, analysis, storage and retrieval (Chen et al., 2012; Sircar, 2009; Wixom et al., 2011).

BI is used in two styles. Traditional style, wherein dedicated IT specialists are employed within organizations to handle large volumes of data and these data are more or less relevant to the financial information and data security of the organization.

Now, with the growing need to process information quickly to aid decision-making based on trends and patterns of events, organizations have now resorted to the second style, which is using agile and intuitive systems to analyze data with great efficiency, reliability, speed and accuracy. The software systems in BI application include Dashboards, Data-mining Reporting, Visualizations, Extract Transfer Load and Online Analytical Processing (OLAP).

The basic flow of BI involves the flow of raw data and its filtration into labelled groups to enable users to access specific information with greater ease in less time. To understand BI better and its application in Business context and specifically in Human Resource (HR) Management, it is necessary for the users (first time especially) to understand the components of BI (Datapine, 2020).

i. **Raw Data**: The first element of a BI solution that supports critical decision-making are the data themselves. A company obtains its data from multiple sources, including sales records, customer satisfaction surveys, profit and loss statements and other data sheets that are stored in databases, enterprise resource planning (ERP) systems and document files in systems. BI allows the system to retrieve data from across databases.

ii. **Database Warehouse**: The database warehouse is a logistics platform connecting different data storage systems in the organization. These data warehouses can be connected based on the likeness of data thus creating a correlation. Like a physical warehouse stores multiple products, the data warehouse stores data, semi-processed and processed information from multiple reliable sources that can be easily stored and accessed by the user.

iii. **Data Access, Analysis and Retrieval**: The data stored in the warehouse are accessed, analyzed for trend study and interpreted into meaningful forms through data mining. Data mining involves classification of data, prediction of values (similar) and affinity or information cluster formation. Data mining helps in visualization of data, exploration of large sets of data and consolidation or minimization of data into meaningful forms before presentation. The data-mining process involves
   a. setting clarity of goals;
   b. collecting data from various reliable sources;
   c. pre-processing data for relevance and reliability;
   d. reducing data into validation and test sets;
   e. classifying and predicting data;
   f. applying commands and algorithms to interpret results;
   g. applying BI models.

iv. **Dashboarding and Reporting of Data**: On completion of the data-mining process, the final information is presented through dashboarding and reporting. The most important step in this stage is to effectively track the inflow of data, monitor the data and report the data to the user.

By setting a dashboard that is customizable, flexible and driven by data, users can identify trends and patterns to effectively predict future insights (Bagale et al., 2021). These insights can be effectively shared within the organization system or to desired stakeholders through data visualizations. Data visualization or visual BI supports quick access of data in visual forms (Baltzan, 2014; Iliinsky & Steele, 2011; Rodeh et al., 2013). This being an emergent technology to meet the ever-dynamic inflow of complex and extensive data, the technology offers the ease of grouping datasets based on statistical data representation and analysis (Alazmi & Alazmi, 2012, p. 297). Data visualization also supports data to convert information into a business viewpoint (Baltzan, 2014).

BI is gaining popularity due to its impeccable capability (Kautish, 2008; Kautish & Thapliyal, 2013) to help knowledge management professionals recognize information trends from data patterns collected that are used to support organizational decision-making. This decision-making incorporates usage of computer technology and analytical algorithms to process raw data of immeasurable quantities into interlinkage chain of information, which can be accessed and retrieved with comparatively less time (Chaudhuri et al., 2011; Chen et al., 2010; Williams, 2011).

Globally competent organizations now follow the technology to lead the organizations' BI and decision-making (Alazmi & Alazmi, 2012). BI has gained popularity in the past years and now with its ability to easily blend with organizational departments have been seen as an innovative step to be adopted in the future (Chaudhuri et al., 2011; Manyika et al., 2011; Russom, 2011).

The purpose of this chapter is to introduce BI as an effective technique to manage high volume of volatile data not only in technological specialized environment but also in allied fields of specialization, including Finance management and HR Management, which deals with extremely sensitive data processing.

4.2.3 Common Myths Relevant to BI

Along with increased acceptance for impeccable accuracy and data privacy, BI also has been misunderstood by commoners who believe

- BI when applied produces astounding transformation of business organizations.
- The techniques of BI are complicated and can substitute domain knowledge and experience in information analysis and information and decision model building.
- BI tools can automatically extract information pattern and create trends and patterns out of them.
- BI is most effective in the scientific and management areas of finance, marketing, logistics and fraud detection only.
- BI is a complex process and way different than the quantitative model-building process.
- BI and data mining are ineffective because the mining process can lead to erosion of data, especially if unused.
- BI is another fad technology hype, which will fade off, taking organizations back to business processes in a traditional way.

To break the above myths, the role and significance of BI in the Industrial Revolution 4.0 (IR 4.0) Era needs to be studied in order to clearly establish how BI as a business information technique can support decision-making in organizations.

4.3 Role and Significance of BI in IR 4.0 Era

4.3.1 Industrial Revolution 4.0

The IR 4.0 marks the recent wave of technological application and digitalization in industrial sector driven by the digitalization, automation and comprehensive networking (Zambon et al., 2019). Past decades have seen remarkable leaps of digital upsurge, which has led to individuals and organizations being heavily dependent on technology for all their needs.

While criticism that excessive technology application leads to reduction of human workforce was heating up, the 2019 pandemic turned the event flow once and for all. With the pandemic causing significant loss of lives and demanding work from home, digitalization of workspaces has become inevitable. With this digitalization, it has been observed (Zambon et al., 2019) to have led to improved production and client management in improved efficacy. Additionally, IR 4.0 has also increased the need to constantly innovate product and process technology and redefine and redesign business models. This drive has in turn inspired non-technical organizations, including legal, political and allied business systems, to adopt the IR 4.0 approach.

BI as a technology supports business activities in innumerable ways. The role of BI systems may be assessed from multiple dimensions. BI is also expected to play a pivotal role in creating sustainable organizations through value-driven changes that may aid an organization in business development, creation and retention of loyal customers, stakeholder management through timely and effective decision-making, new market creation and diversification, and extreme possibilities of innovation (Chaudhary, 2004; Olszak & Ziemba, 2004; Reinschmidt & Francoise, 2002).

BI as a technology is capable of transforming organizational data in a more relatable and meaningful manner. BI professionals use various software and apt technologies to meet the organizational needs of gathering, storing, analyzing and interpreting data. However, it is to be noted that the pivotal purpose of these systems is to reach into systematic and strategically driven decision-making based on facts and information. Organizations collecting data from a large number of sources maintain a data warehouse to store information for the creation of a BI system.

BI being a reliable technology to handle large volumes of data that are confidential and volatile needs to ensure the data so collected and stored in data warehouses are

- handled by experts with credentials to access data;
- assessed for data capacity and validity;
- appropriate data retention process flow is set.

All of the above actions need tools and techniques to be applied based on the requirements of organizations and the nature of data. The tools in BI include

- *Business Planning and Process Re-engineering*
- *Business Performance measurement and performance assessments*
- *Scorecards*
- *Competitor and market analysis systems*
- *Decision support system (DSS) and data forecasting*
- *Data mining, text mining, data and document warehousing and management*
- *Enterprise management system (EMS) and Information System (EIS)*
- *Online Analytical Processing (OLAP) and multidimensional analysis*
- *Associative Query Logic (AQL) and Customer Relationship Management (CRM)*
- *Marketing and HR Analysis*
- *Finance and budgeting*
- *Management Information Systems (MIS) and Geographic Information Systems (GIS)*
- *Research and Statistics Data Analysis*
- *Logistics and Supply Chain Management*
- *Trend Analysis*
- *End-user Query and Reporting*

BI is a thoughtful and planned application of interventions into business environments involving the collection of raw data, data cleaning and transformation/transfer of data into data warehouses and analysis. An effectively applied BI system can integrate information to derive trends from them and forecast the future scope of business. BI transforms raw data into valuable information supporting decision-making. Organizational decision makers need to perceive the role and significance of BI as a predominant element to support effective decision-making through the management of knowledge and sharing of reliable information, open access and transparency of communication, analytic process flow to lead organizations to solutions that are easily trackable thus creating effective strategic decision-making as illustrated in Figure 4.1.

BI has been assumed to aide strategic decision-making at all levels of organizational management and development regardless of the challenges and complexities in the structuralizing (Olszak & Ziemba, 2003). The strategic application of BI ensures the preparation of organizational reports by effective analysis of past data, profitability trends to predict future results relevant to marketing, finance, expenditure and capital management, etc., thus allowing optimization of future decisions and actions taken by the organization (Figure 4.2).

At an operational level, BI applications are implemented to perform adhoc analysis using data evaluation and pattern study to lead to data-driven decision-making in organizational performance management involving (Hsu, 2004; Olszak & Ziemba, 2003)

- reviewing of costs and revenues from financial reports;
- income and expenditure analyses;

Figure 4.1 Generic process flow of data in business intelligence.

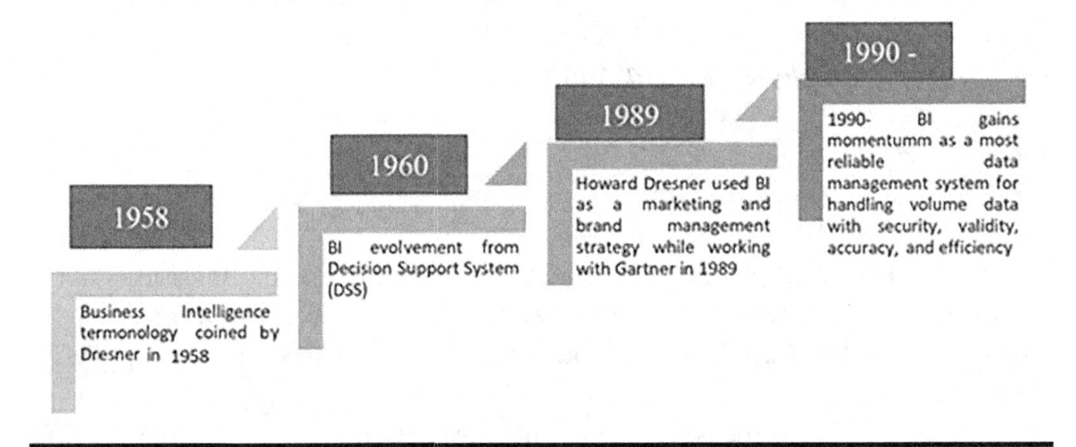

Figure 4.2 Evolution of business intelligence.

- balance sheet and profitability analyses;
- financial market analysis and prediction;
- analysis of sales, profitability, sales targets and order management;
- competitor and stakeholder management;
- customer creation, analysis and retention through customer touch point and buying behaviour, management identification of production and logistics 'bottlenecks';
- observation of employee performance trends and wage data management;
- employee turnover analysis and management of employee's personal data; and
- assessment or organizational discrepancies and frauds (Olszak & Ziemba, 2006).

4.3.2 Benefits of BI in IR 4.0

With the 2019 pandemic disrupting work environments, BI has gained more importance due to its effectiveness in business decision support. BI enables the supply of reliable information (Xu et al., 2018) to industry with ease of access thus eliminating speculation of what could be done to improve the performance of organizations. The organizations that need to sustain and survive amidst competition need to respond to ever-dynamic demands and consumer responses in the least amount of time. To enable this need, inter-departmental coordination and communication need to improve. BI helps to expedite organizational decision-making and often ensures that the strategic point of view and the sustainability goals are given importance without compromising superior performance. Through the pandemic, some of the benefits provided by BI apart from validity, visibility, accuracy, easy adaptation and reliability included

- quick conversion of business information through analytical and artificial intelligence thus resolving many issues, even critical logistics and customer retention;
- organizations had better support to identify discrepancies in business process flow and rectify them within the shortest time span;
- control of large volume of data from virtual environments was made possible through cloud data warehouses managed by specialists;

- analysis of potential customers, creation of loyalty and retention of loyal customers was made possible;
- management and verification of click-stream data in e-commerce platforms;
- increased logistics and reduced product warrant deficiency support;
- reduce fraud and reduction of equipment downtime through proactive and predictive interventions, etc.

Yet another major advancement driven by BI amidst pandemic and increased business challenges faced by organizations is the reduction of cost to organizations. Owing to the product and process redundance of products and processes during the pandemic, the organizations geared by IR 4.0 has been facing a reverse challenge wherein pull strategy management by BI has to be implemented as a push strategy to save the organization from closing down. How BI has supported infrastructure cost is worth analyzing and understanding as a step to aid in disaster management for the future.

First, BI by eliminating redundant data and duplication of data during the extraction process reduces the infrastructure cost borne by organizations. 3M met this challenge to save its multi-million worth data warehouse platform using BI (Watson et al., 2004, p. 209). BI through ease of data storage and retrieval saves time for end users, may it be suppliers or consumer data experts. Efficient data-driven forecast can be made to answer the question of what has happened to reach the present scenario based on which future actions can be planned and implemented through data-driven forecasting. In future, the mature application of BI can support strategic decision to tap unattended, improve branding and market positioning and be a sustainable organization through customer-centric approach.

BI is now growing as a prominent and preferred data management system with its ability to blend itself into systems creating wide interest, including business performance management and re-engineering and real-time BI.

As a part of IR 4.0, when organizations moved to powerful means, Internet of Things for all, the Web-based systems enabled data access from anytime and anywhere. This in turn has allowed easy tracking of resources in the warehouses of organization and quicker order placement based on consumer demand prediction, all done through effective BI reports and decision-making.

4.4 Scope and Application of BI in HR Environments

BI is a comprehensive field combining technology, business process components, HR skills, strategic planning and tactful implementation of techniques. The components of BI include collection, cleaning, storing and sorting data into warehouses enabling easy retrieval. The technologies and skillful intervention is expected to effectively support data querying, report generation and presentation of information in a more meaningful way to aid in decision-making. The data sorted and stored in the data warehouses are sorted into specific data marts that relate to specific departments of an organization, which also includes the HR department.

IR 4.0 organizations face multiple challenges and among these challenges lie opportunities that need to be seized. The most important role of BI is to support these organizations to seize these untapped opportunities by rightly providing the necessary information for effective and timely decision-making. With technological advancements, and wide usage of Internet technology, came the need to obtain accurate information in the form of reports to meet the rising demand of customers organizations. For this there is a need to quicken the decision-making process and with the increased information overload, even highly competitive organizations are feeling the pressure to respond. With competitors watching over the shoulder, organizations are heavily dependent on BI to take fact-based strategic decisions. By the year 2020, the data generated yearly by organizations is expected to touch up to 35 zettabytes (1 zettabyte=1 billion terabytes, 1 terabyte=1,000 gigabytes) (International Data Corporation, May 2010).

This data if responsibly managed through BI application will considerably reduce the data maintenance cost and optimize HR and related business operations. BI is a preferred solution not as a transactional process but as a one-time investment that can save the time and cost with the benefit of opportunity gains (Davenport, 2006).

With the globalization and run to excel, the challenge of harnessing and interpreting information relevant to product, process, transactions, inventory, customer, transactions, competitor and business data has been vested on BI (Chen et al., 2010, p. 201).

Organizations that have been applying BI since years have been reporting efficient data management and improved process and decision-making with the implementation (Chaudhuri et al., 2011; Turban et al., 2011). Chen and colleagues (2012) have remarked BI as an area of study in business and academics alike.

Business environments are highly dynamic and prone to constant change. With the rise in competition and digitalization of process systems in IR 4.0 (Friedman, 2004), decision-making has become increasingly challenging and in addition the availability of low-cost labour and rise in consumer demand in developed and semi-developed countries has pushed organizations to go global with increased international opportunities being sought. This step has increased the pressure on 'HR environments to manage recruitment and manage diverse work force accommodating varied time zones, virtual work environments, cultural differences and expertise'. Realizing 60%–70% of managerial cost is attributed to managing people power within organizations, business systems and organizations are taking drastic steps to reform and redesign organizational structures to ensure informed decisions are made based on knowledge and data. It has been established that the organizations that successfully implemented BI to manage their data and decision support systems have been able to take informed strategic decisions quickly and with greater precision. This is because of the transparency and ease of access to functional departments and their data.

The role of BI in the HR system can be better explained through an example. The State of California set two objectives (http://www.dpa.ca.gov/hr-mod/accomplishments-and-goals/mission-statement-goals-andobjectives.htm) as a part of improving and increasing performance efficacy at the workplace:

Within state services, employee wellness and development are often a formality, which is not seriously adhered to due to the expense it creates to organizations. The objectives set by the study were to

- Ensure supervisors and managers in the organizations acquire, implement and apply principles necessary to nurture increased performance of employees in the workplace. Currently performance appraisals were performed occasionally even though they were supposed to be done annually. These appraisal reports are necessary to assess the current competencies of employees and learning gaps to ensure proper training steps are imparted as timely training and development is an inevitable need for effective and increased performance.
- To ensure the supervisors and managers conduct the performance appraisal in the most effective manner and develop the skills to provide constructive and effective feedback.

In the above research conducted the data were collected and evaluated using BI to support decision-making and implement corrective measures in the shortest time span.

Many other corporates have resorted to BI to perform various HR functions ranging from candidate recruitment, filtering based on skills and competencies sought by the organization, data management and performance report generation. An important result attained by this endeavour is selection of the right person for the right job and the retention of the employees based on the performance evaluations. Further, as represented in Figure 4.3, BI in HR systems also supports effective analysis of organizational performance through a balance scorecard wherein all information and goals that could impact organizational performance is laid clearly through a well-developed business plan. The plans are then communicated to the employees through transparency systems, trainings and orientation programmes. Following the communication, the employee performance is effectively linked to rewards that motivate the employees to do their best to achieve organizational goals. Analysis of employee performance

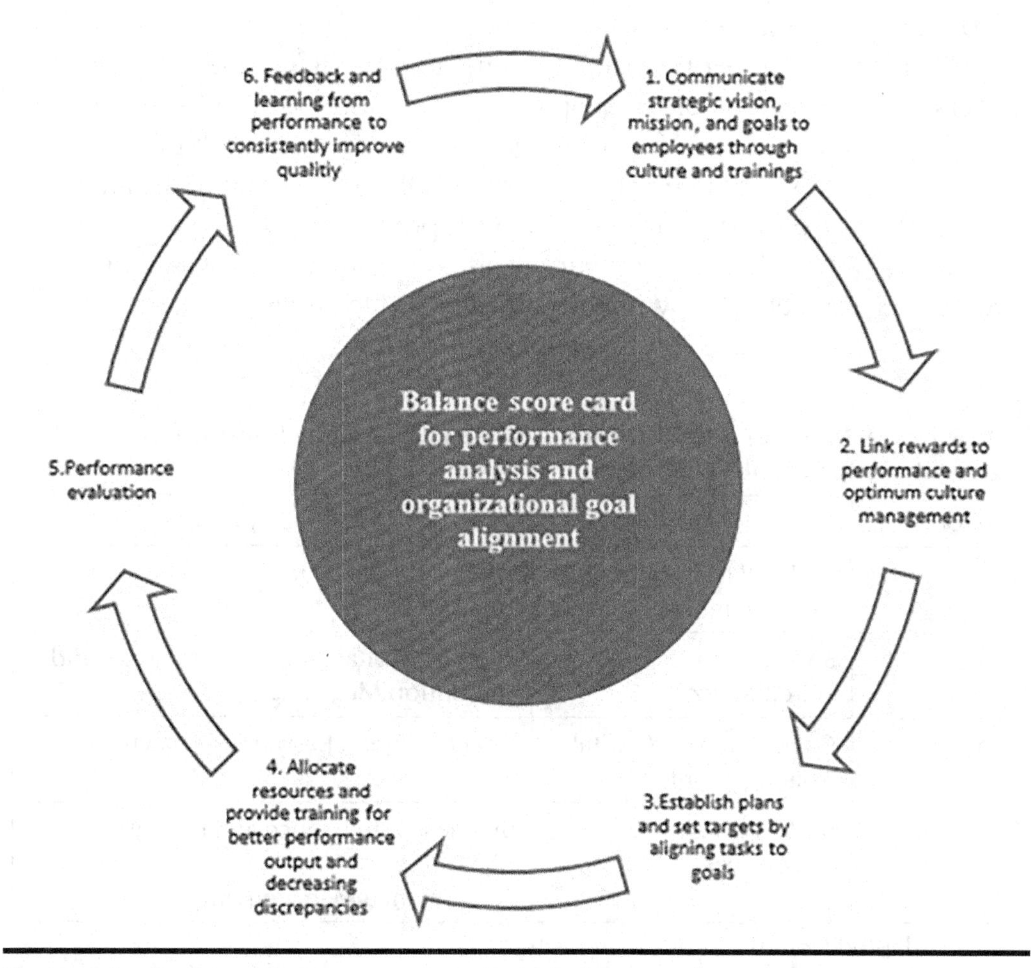

Figure 4.3 Performance management strategic framework.

allows assessment of how much of the organizational goals has been achieved through employee performance and any discrepancies in employee performance are corrected through constructive feedback and training that form the learning support. The Balance Score Card is an effective task carried out by BI, thus making the performance evaluation, reporting and presentation quicker and more accurate. This is one typical example of how BI-driven information supports strategic HR decisions and practices.

BI popularity has been growing faster than predicted as evidenced in the report released in April 2010 (Gartner, April 2010). BI process systems: analytic techniques and HR software sales revenue crossed a $9.3 billion in 2009, which is a steady 4.2% increase from the previous year, which generated $8.9 billion.

BI platforms captured 64.2% market share globally with $5.98 billion of BI software revenue. A total of $1.94 billion, which accounted to 20.8% market share, was taken by corporate performance management suites and performance management systems and analytical software with $1.40 billion, i.e., 15% of the overall market share.

With the tremendous opportunities generated by BI, a large number of mergers and acquisitions of BI companies by globally powerful software giants have taken place. Profits have been coming in as a result of these acquisitions since we've been able to provide BI services, solutions, and products that have been smashed by service providers.

Table 4.1 summarizes the software vendors and their products, which have been most preferred by organizations to reduce their HR-related tasks, data collection, reporting and performance tracking.

Table 4.1 HR Modules and BI & Data Analytics Offered by Software Organizations for Giant Vendors

Vendor	*HR Module*	*BI & Data Analytics*
SAP	SAP ERP Human Capital Management	Workforce Analytics
SAS Institute	SAS Human Capital Intelligence	Human Capital Predictive Analytics and Retention Modeling
Oracle	Oracle Human Capital Management	Oracle Human Resources Analytics
IBM	IBM Cognos – Human Resources	Business Intelligence and Human Resource Performance Management

Source: Adapted from the vendors' websites.

The SAP ERP Workforce Analytics Software (*SAP ERP Workforce Analytics Software*, 2019) has been claiming the following features through its software:

i. Workforce planning by manpower planning
ii. Turnover rate assessment
iii. Workforce cost prediction
iv. Payroll and employee benefit evaluation and analysis
v. Performance evaluation and talent management analysis
vi. Balanced scorecard framework to integrate employee management-by-objective (MBO) to align employee efforts to corporate goals.

The ***SAS Human Capital Intelligence Software*** (*SAS Human Capital Intelligence Software*, 2020) claims to have the following features:

i. Employee turnover prediction with risk analysis of voluntary termination
ii. Voluntary termination analysis
iii. Category-wise job risk identification
iv. Employee turnover and risk involved leading to turnover.

The ***Oracle Human Resources Analytics*** software (*Oracle Human Resource Analytics*, 2020) features are as follows:

i. Targeted workforce development and workforce insight monitoring focusing job and delivery methods of organizations
ii. Compensation and work force cost planning and management to assess cost-effectiveness of employee compensation techniques
iii. HR Performance Assessment with Leave and Absence tracking and reporting
iv. Employee productivity tracking and benchmarking standard workforce processes. Compare the measurements with external benchmarks and internal operating thresholds
v. Workforce process analytics and support of core HR processes, such as payroll, time management and compensation
vi. Talent management analytics and training support
vii. Succession planning programmes
viii. Employee goals and MBO alignment

The IBM Cognos Business Intelligence software (*IBM Cognos Business Analytics*, 2020) includes the following features pertaining to the core HR areas:

i. HR staffing, training and development
ii. Talent management, compensation and succession
iii. Compensation cost management for improved performance.

Although BI software is creating substantial gains to organizations, the full advantage of BI in HR management area is still minimal. The reason for this is the perspective of organizations that view HRs as a cost centre and never an asset; BI can reduce the cost involved in HR systems, thus bringing in more gains and exceptional success leaps to the organizations. To realize this objective there is a great demand for specialists who can drive the BI systems towards desired goals. The increased interest in BI has fuelled the demand for knowledge professionals and information management experts with increased critical analysis skills.

Despite the rising demand, there is a shortage of employees in this area of expertise. Studies by the McKinsey Global Institute revealed a gap of 50%–60% in the supply and demand of experts in data analysis and management areas, which equals approximately 140,000–190,000 vacant positions, suggesting a shortage of serious analytical talent (Manyika et al., 2011; Russom, 2011). This has further increased the scope for adding more programmes relevant to BI in universities to meet the employee shortages in IR 4.0 era (Connolly, 2012; Sircar, 2009; Wixom et al., 2011). Connolly (2012) suggests BI as an effective technique to evaluate and sort massive data to make sound decisions creating business value.

A recent BI Summit evaluated the increasing presence of BI in academic research as an effective mode of data pattern management (Wixom et al., 2011). Recent studies, due to the rising need of processing global information wave, highlight the necessity to educate new generation on the use and scope of BI in business management (Chen et al., 2012; Chiang et al., 2012). Corporate employers and academicians also strongly agree to the need for exposing students to BI concepts and application as a part of preparing them for promising opportunities in the coming years (Connolly, 2012; Conway & Vasseur, 2009; Sircar, 2009; Wixom et al., 2011).

BI due to its growing importance has gained a place in the IS 2010 curriculum for undergraduate degree programmes and has also been included as an important topic under the Data and Information Management stream (Topi et al., 2010). Given the scope of managing the rather tricky flow of information from multiple sources, the professionals in the role of managing data are expected to understand and apply BI concepts, theories and techniques into workplace scenarios on graduation. The success of BI in business context also requires organizations to carefully integrate the 'learning by doing' methodology through on-site projects and internships to gain skills and competencies to apply the theories learnt (Chen et al., 2012). The more the students are guided to apply knowledge, the more they end up being successful practitioners of BI (Piyayodilokchai et al., 2013). Wixom and colleagues (2011) also stress the need to get learners of BI to critically interpret and analyze business scenarios and issues to amicably resolve the critical issues they find.

The beauty of BI application lies in its efficiency in operation by blending itself well into the managerial process systems of organizations. The raw data conversion with maximum speed and accuracy delivers key information to business users. This efficiency is achieved by the standards and operating procedures set. The integration of data from various reliable sources empowers users to be self-sufficient in analyzing data to make faster and better decisions.

4.4.1 Application of BI into HR Environments: The BI Infrastructure

Global organizations can gain a competitive advantage if a well-designed and implemented BI structure is in place. Implementation of BI to transform organizations is no easy task. This requires an effective culture of transformation imbibed in the organization. The following prerequisites are necessary to initiate and fuel the transformation of organizations into sustainable global institutions:

- Research and development capacity of the organization
- Extreme co-operation among the users involved: management, IT specialists
- Transparency and information sharing
- Organizational ability to interpret information properly.

The right model of BI suiting an organization requires identification of role of users in the BI process because the success of BI systems depends on how effectively the users

- identify the model and its working;
- how effectively the system is monitored and reports analyzed;
- how best the reports are utilized for strategic decision-making purposes.

Keeping the above requirements in mind, two major stages of implementation are suggested: (Dresner et al., 2002)

a. BI creation, which consists of definition of the BI, the role and determination of BI system development strategies; identification and processing of data; selection of appropriate selection of BI tools; implementation of BI; and exploring futuristic application scope for BI.
b. BI 'consumption' deals with the end-user application and the steps mainly include logistic analyses; data access and monitoring; development of alternative decision alternatives; and transformation of organizational performance.

4.4.2 Operational Set of BI in Organizations for Strategic Decision-Making

BI is essentially a step-by-step ***implementation*** of technology to achieve an end result in a decision-making role. The process ranges from setting up the automation system to the end-user use of data in decision-making. There are primarily three phases in the automation and implementation of BI systems in organizations:

a. **Automation of BI Framework in Organization**: Here, the organization is expected to list out the process structures within the organization and identify the departments that need to be tracked by the BI system. As a start process, the reporting systems in the organization include the stock-reporting system, payroll and employee data sheets, accounting and bookkeeping systems collectively known as the ERP system, automation of the CRM and the HR management system.
b. **Data Management**: The data management stage is one of the most complex yet crucial stage in data automation, and data scattered in multiple resources need to be collected, classified and sorted and the process can lead to reduction of quality, loss of data, duplication of data before it is properly warehoused and metadata is created.
c. **Business Intelligence Phase**: In this phase the execution of BI interventions planned starts off.

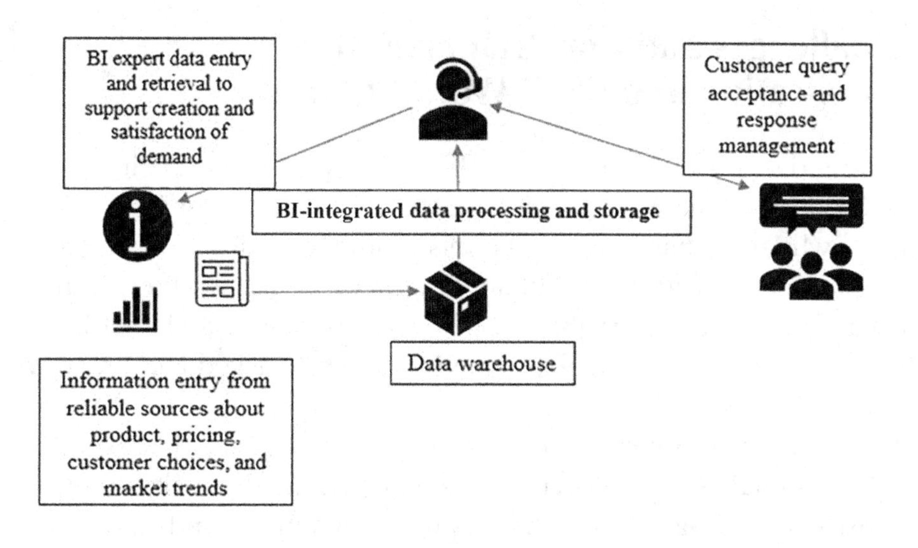

Figure 4.4 Example of BI process flow.

With the application of BI, as represented in Figure 4.4, information fed in the form of product or process addition, marketing strategies, customer need analysis and support data input through the product, marketing and customer demographic databases is stored in metadata units within the data warehouse. BI as a system provides reliable information in real time across the enterprise with better validity of data. It also allows data visualization through meaningful dashboards allowing effective decision-making. The BI system from the backend provides analysis of how the changes fed into the data system impact the business by analyzing the revenue stream fluctuation with a new price, increase in customer creation with added marketing strategies in place and customer reactions to price change of a product.

d. **BI Project Testing**: Once the BI platform is set and automation of information is done, the organization conducts the BI system testing for data quality, model testing, integration testing, LAP cube testing, regression testing, data completeness, scalability and performance, user acceptance testing, report testing and data accuracy and reliability.

The success of BI system depends on how well the implemented system fares in the above tests. Effective BI integration and tests build trust and confidence in users on the reliability of the information that will serve crucial strategic the BI data reports generated.

4.5 Challenges and Constraints in BI Application into Rigid Work Culture

After having discussed the benefits of BI and its seamless scope in the areas of business and HR management, it is necessary to think despite these researches why there has been constraints in implementing BI successfully into work environments. The successful application of BI into work environments requires a progressive culture that is built on transparency and coordination. The following are some of the reasons why BI is not invited into organizations:

i. BI is costly as it involves technology and process cost involved, which could be challenging for typically small- and medium-sized enterprises.
ii. Constriction of data warehouses is time consuming and overly complex effort as it requires collection of data from multiple sources, and during the implementation, there is risk of losing data validity and reliability.
iii. Testing of data warehouses and its implementation success requires time, and in highly competitive organizations time is a constraint.
iv. Lack of proper understanding of the benefits and uses of BI in business and other areas of management.
v. Lack of professional skills among employees to handle high-tech data-driven systems.
vi. Lack of reusability, audit trails and quality testing methodology make the quality questionable.
vii. Difficulty in convincing stakeholders and investors about the benefit of BI systems and the time to see the success of the system as the scope, time and budget are equally important. On time delivery demand can be a pressure to professionals, which may lead to failure of BI system or loss of quality.
viii. Need to train existing employees to adapt to a new system may lead to challenges of resistance to change, which requires appropriate intervention to manage change.

4.6 Model to Apply BI in Organizations to Create Sustainable Organizations with the Proposed Methodology and Process Flow

To successfully implement the BI system into an organization, it is first necessary to understand the vision of the organization. A sustainability driven organization is identified by the developmental efforts taken by them to build a strong culture and innovative work systems. BI systems can be implemented in organizations by adopting a sustainable BI strategy, which involves the following efforts:

i. Proper understanding of the vision of the organization

   The BI system must be well aligned to the sustainability goals of the organization and the entire management, and employees must be aware that the step taken towards sustainability is not a quick process and would require investment of resources and time.

   The key IT specialists heading the innovation drive of the organization must carefully plan the information resources needed to implement step-by-step installation of BI system, which becomes a crucial part of the business strategy (Chaudhary, 2004).

   The success and sustainability of an organization is achieved by the attainment of three tangents: environmental sustainability, social value sustainability and economical sustainability.

   While environmental sustainability is characterized by low carbon footprints, reduced pollution and use of renewable energy, social value formation for sustainability includes establishment of equality and well-being in the community. Equality in labour, life standards, health and safety are also a desired outcome from sustainability. Organizational sustainability also is secured by steady profits, timely return on investment financial stability and long-term business stability.

   It is necessary to understand that BI is not a passive process and is an effort to ensure the longevity of the organization. Hence, even the data collection and recording in the BI systems is not expected to be a short run or callous effort. As a part of this venture, it is necessary to understand customer expectations and watch the trends of customer preferences, which will help predict the future demand of the products and services. The pivotal responsibilities of business enterprises wherein BI is implemented may include (Kalakota & Robinson, 1999; Liautaud & Hammond, 2002; Rasmussen, Goldy, & Solli, 2002) the following initiatives:

   - Transition from a probability-based decision-making to a data-driven decision-making using facts and balance score cards and other analytical approaches.
   - Align the strategic decisions with customer and supplier behaviour.
   - Implementation of standards of quality, including measures for repetitive and regular quality.

- Rapid investigation and redressal of frauds and business process discrepancies.
- Reduction of time frame for data processing, reporting and decision-making.
- Analyze information and decrease the number of participants who are involved in analyzing and processing of information; and

Currently BI has been successfully implemented in banks and finance-related/insurance sector organizations followed by the telecommunication sector and transportation organizations. Production-based companies and logistics organizations have also found benefits with BI application into process flow and decision-making within organizations.

General trends of BI implementation in organizations show the adoption of a top-down approach of management and decision-making. It is also necessary to decide whether BI is to be implemented in one department or throughout the organization.

Hence, the objective is more to evaluate the goals and vision of business even if it is challenging.

ii. Preparation of appropriate data collection system for installation the BI System.

This stage is characterized by automation of all organizational data, which includes invoices, HR and financial records, customer queries and satisfaction records, production and supply chain process flows. It is also necessary to analyze the information collected for reliability, validity and security by following the standard procedures of (Błotnicki & Wawrzynek, 2006):

a. Filter the important data from the passive data.
b. Generate links and find relationship between similar data from multiple sources.
c. Establish the logical structure of system-generated data.
d. Find the full scope of the information system to identify loops of error and take necessary precautions to mend the system to meet the quality requirements.
e. While data mining is conducted, text processing and wrappers must be implemented; text-processing methods may prove useful to increase effectiveness (Poul, Gautman & Balint, 2003).
f. Clean your data using effective approaches to increase reliability and support in decision-making.

iii. Develop a data dictionary

It is necessary to decide whether the BI application is restricted to one department or is being done organization wide. In this process, it is worth developing data dictionary to set metadata warehouses to easily store and retrieve data.

iv. Determine the Key Performance Indicators (KPIs)

BI is a serious step that involves investment of cost. Resources and time and hence the stage-by-stage success of the implementation process must be measured. This can be done by the management by setting KPIs that are objective oriented and measurable. Yet, another benefit of the KPIs is that they allow early detection of deviances or discrepancies in the application process providing opportunity for rectification and realignment of organizational vision and goals to the BI process.

v. Data cleaning and data mining for retention of reliable data

The data requirements must be clearly articulated to support and ease the data-mining process. Sufficient quality and quantity of data must be made available for the data-mining process to be effective.

vi. Selection of appropriate BI Software and Tools

The choice of BI tools and techniques to be applied is a tricky challenge. Often all organizations get stuck in this process due to the push strategy adopted by the software providers. BI software providers have amassed huge profits in the last decade by supporting organizations in the BI system implementation with modules and process systems. Companies offer a wide range of software starting from simple data collection, sorting and analysis to sophisticated BI platforms. Factors, such as reliability, cost, functionality, support from solutions and organizational compatibility, must be focused before deciding the right BI software. Organization's goals and informational needs pertaining to market, and customers needs have to be considered, and for this the products supporting BI may be found in different IT segments (Ilczew, 2006). Along with the software itself, the necessity to align the BI system to customer needs and supplier demands must be checked for. Software giants also provide customized solutions at a premium price.

vii. Experts to manage the implementation of BI Systems

Great care is taken while implementing BI systems stage by stage. However, in the creation of a customized BI the time taken for installation is more. This time is utilized for designing user interfaces and checking the system for logical consistency. Designing of BI also involves the crucial process of constructing data warehouse, which proves to be a repository for data analysis, and of a storage system for the BI system (Inmon, 1992). This process must be in alignment to the following (Błotnicki & Wawrzynek, 2006; Hackathorn, 1998):

a. Setting the scope of organizational data and the purpose of the data
b. Defining the relation between data
c. Creating a data model as that of a 'a star' or 'snowflake' that supports implementation of the system including OLAP.
d. Data import systems for real-time data must be set for (Meyer, 2001) importing all data, processing of data and reporting. Apart from system-generated standard report, the BI system must also give freedom to develop customized additional reports. It is assumed that in near future organizations must settle to multidimensional analyses (Rasmussen et al., 2002), where BI supports OLAP modules allowing users to mine and retrieve data through data mining providing extensive information relevant to demography, geography, customer preferences and perspectives (Grossman, 1998; Perkowitz & Etzioni, 1999).
e. Depending on the needs of the organization, BI systems must also be able to execute the following (Olszak & Ziemba, 2006):
   i. Segmenting and profiling customers and assessing customers' profitability and value creation.
   ii. Monitoring consumer loyalty and returns increasing efforts to customer retention.
   iii. Identifying discrepancies and frauds.
   iv. An appropriate interface must also be designed. Organizations may even take the support of eternal service providers for this stage (Ilczew, 2006).

All of these are no simple tasks and need the hands of skilled professionals to tactfully implement the stage-wise process framework. It is also advised to have dedicated BI personnel who can lead a team of management and information technology to run the BI system implementation for greater effectiveness and reduced flaws.

4.7 Discovering New Informational Flow and the Way Forward

The stage where once BI implementation is done, and new data are expected is of great importance. Globalization has led to a downpour of massive information wave on organizations wherein the new implemented

BI system is expected to cast a new light on the nature and significance of each type of information to the organization and its decision-making process. Ensuring effective approaches of data mining and system designing, the developed BI system must serve as an effective tool of evaluation to the BI application (Dresner et al., 2002).

Organizational performance changes involve search for advanced forms of support and or outsourcing to lead to new partners and unexplored markets (Linoff & Berty, 2002; Srivastava, 2003). This accomplishment does not guarantee the resolution of all the issues of an organization. Being of iterative nature, BI systems keep creating loops that urge information analysis and adjustments. As a part of completion of BI implementation, the acceptance of BI system by the end users is particularly important. This will ensure the system can perform the tasks expected based on test reports and trial runs. The testing process is an ongoing activity because defects and discrepancies must always be detected earlier and corrected to avoid exponentially increasing costs.

4.8 Future Scope of BI

The BI techniques are already prevalent in data-driven industries, such as banking and insurance. Research reveals that more than 83% of government organizations in developed countries have transitioned to BI systems to support data-driven operations and decision-making. While BI implementation has been preferred and applied increasingly in global industries, the scope for BI to be applied into digitally naïve sectors are opportunities to be tapped. The following are some areas where BI and analytics approaches that will be expected to change the future of organizational decision-making (Columbus, 2018):

a. **Cloud Analytics**: BI applications will soon be available as cloud support systems as with the pandemic in 2019–2020; organizations have understood the necessity of anytime and anywhere reporting and information access, which has made physical information systems and data access station impractical.
b. **Collaborative BI**: BI combined with social media, and teleworking platforms will be used for virtual work environments and strategic decision-making purpose.
c. **Artificial Intelligence and Machine Learning**: Gartner's report indicate the rise in role of machine learning and AI to take over tasks previously dealt by human intelligence. This also includes customer

preference analysis. Healthcare statistics generation, chatbot management and more.

d. **BI-Embedded Systems**: In future, BI-embedded systems where BI technology is integrated wholly or partially into other business applications will come in place, thus enhancing and extending the scope of BI from its reporting functionality to much more.

4.9 Conclusion

The future of BI is yet to be seen owing to the wide scope it holds depending on how well the technology can be made useful in all fields of human life and work. To develop the scope of BI, organizations need to automate the information systems of the organizations thus initiating the first stage of BI implementation. As the benefit of BI is not to be restricted to just financial institutions and industries, BI systems need to be made flexible and organizations need to make most out of the BI systems. Although the initial timeline will be challenging with appropriate leadership and time-based training, the culture of the organizations can be transformed into a conducive development-oriented workspace. BI solutions must be scalable by providing increased flexibility to allow for expansion of the system, which will provide a greater chance for the effective management of an enterprise. With right leadership, culture and BI experts, organizational sustainability is more a realization in the coming years for global IR.4.0 organizations.

References

Alazmi, A. R. R., & Alazmi, A. A. R. (2012). Data mining and visualization of large databases. *International Journal of Computer Science and Security*, 6(5), 295–314.

Bagale, G. S., Vandadi, V. R., Singh, D. et al. (2021). Small and medium-sized enterprises' contribution in digital technology. *Annals of Operations Research*. Doi: 10.1007/s10479-021-04235-5.

Baltzan, P. (2014). *Business Driven Information Systems*. (4th ed.). New York: McGraw-Hill.

Błotnicki, A., & Wawrzynek, Ł. (2006). Od porządkowania danych do Business Intelligence – jak uświadomiona wiedza staje się elementem konkurencyjności organizacji. [From sorting data to business intelligence - How does conscious knowledge become an element of corporate competitiveness?] In A. Binsztok, & K. Perechuda (Eds.), Koncepcje, modele i metody zarządzania informacja i wiedza. Wrocław: AE.

Business Intelligence Articles, Guides & News | datapine. (2020). Datapine. https://www.datapine.com/articles/.

Chaudhary, S. (2004). Management factors for strategic BI success. In *Business Intelligence in Digital Economy. Opportunities, Limitations and Risks*. Hershey, PA: IDEA Group Publishing.

Chaudhuri, S., Dayal, U., & Narasayya, V. (2011). An overview of business intelligence technology. *Communications of the ACM*, 54(8), 88–98.

Chen, H., Chiang, R. H. L., & Storey, V. C. (2010). Business intelligence research. *MIS Quarterly*, 34(1), 201–203.

Chen, H., Chiang, R. H. L., & Storey, V. C. (2012). Business intelligence and analytics: From big data to big impact. *MIS Quarterly*, 36(4), 1165–1188.

Chiang, R. H. L., Goes, P., & Stohr, E. A. (2012). Business intelligence and analytics education and program development: A unique opportunity for the information systems discipline. *ACM Transactions on Management Information Systems*, 3(3), 1–30.

Columbus, L. (2018, June 8). The state of business intelligence, 2018. Forbes. https://www.forbes.com/sites/louiscolumbus/2018/06/08/the-state-of-business-intelligence-2018/?sh=25a299577828.

Connolly, D. (2012). Why b-schools should teach business intelligence. Bloomberg Businessweek. Retrieved from http://www.businessweek.com/articles/2012-04-23/why-b-schools-should-teach-businessintelligence.

Conway, M., & Vasseur, G. (2009). The new imperative for business schools. *Business Intelligence Journal*, 14(3), 13–17.

Davenport, Thomas H., & Harris, Jeanne G. (2007, March 6). Competing on analytics: The new science of winning. Harvard Business Press, 218 pages.

Dresner, H. J., Buytendijk, F., Linden, A., Friedman, T., Strange, K. H., Knox, M., & Camn, M. (2002). The business intelligence center: An essential business strategy. Gartner research. *Strategic Analysis Report*, 29.

Friedman, T. L. (2004). *The World is Flat*. New York: Farrar, Straus and Giroux.

Gartner. (2010 April). Market share: Business intelligence, analytics and performance management software worldwide, 2009 by Dan Sommer and Bhavish Sood, April 2010.

Grossman, R. (1998). Supporting the data mining process with the next generation data mining systems. *Enterprise System Journal*.

Grossmann, W., & Rinderle-Ma, S. (2015). Chapter 1: Introduction. In *Fundamentals of Business Intelligence* (pp. 1–15). Springer. Doi: 10.1007/978-3-662-46531-8.

Harvard Business School. (2016, June 21). 10 principles of the new business intelligence. *Harvard Business Review*. https://hbr.org/2008/12/10-principles-of-the-new-busin.

Harvard Business School. (2021, January 20). Business intelligence vs. business analytics - Harvard business analytics program. HU-CBA. https://analytics.hbs.edu/blog/business-intelligence-vs-business-analytics/.

Hsu, J. (2004). Data mining and business intelligence: Tools, technology and applications. In Raisinghani, M. (Ed.), *Business Intelligence in the Digital Economy*. London: Idea Group Publishing.

IBM Cognos Business Analytics. (2020). IBM software. https://www.ibm.com/ae-en/products/cognos-analytics?p1=Search&p4=43700062394531433&p5=p&gclid=EAIaIQobChMI3O-Dh_Kb8AIVUNPtCh1L-AGMEAAYASAAEgJIO_D_BwE&gclsrc=aw.dsJournal.

Ilczew, P. (2006). Podejście biznesowe do wdrażania Business Inteligence. [Business approach to implementing business intelligence.] In J. Kisielnicki (Ed.), Informatyka w globalnym świecie. Warszawa: Wydawnictwo PJWST.

Iliinsky, N., & Steele, J. (2011). *Designing Data Visualizations.* Sebastopol, CA: O'Reilly Media.

Inmon, W. H. (1992). Building the data warehouse. New York: J. Wiley

International Data Corporations (IDC). (2010, May). The digital universe decade – are you ready? retrieved from https://www.bitpipe.com/detail/RES/1287663101_75.html

Kalakota, R., & Robinson, M. (1999). *E-business: Roadmap for Success.* Reading, MA: Addison-Wesley.

Kautish, S. (2008). Online banking: A paradigm shift. *E-Business, ICFAI Publication, Hyderabad,* 9(10), 54–59.

Kautish, S., & Thapliyal, M.P. (2013). Design of new architecture for model management systems using knowledge sharing concept. *International Journal of Computer Applications,* 62(11), pp 27–31.

Liautaud, B., & Hammond, M. (2002). *E-business Intelligence. Turning Information into Knowledge into Profit.* New York: McGraw-Hill.

Linoff, G. S. & Berry, M. J. A. (2002). Mining the Web: Transforming customer data into customer value. New York: J. Wiley.

Luhn, H. P. (1958). A business intelligence system. *IBM Journal of Research and Development,* 2(4), 314–319.

Manyika, J., Chui, M., Brown, B., Bughin, J., Dobbs, R., Roxburgh, C., & Hung Byers, A. (2011). *Big Data: The Next Frontier for Innovation, Competition, and Productivity.* McKinsey Global Institute: McKinsey and Company. Retrieved February 19, 2021 from http://www.mckinsey.com/Insights/MGI/Research/Technology_and_Innovation/Big_data_The_next_frontier_for_innovation

Meyer, S. R. (2001, June 1). Which ETL tool is right for you? DM Review Magazine.

Morris, Steven. (2018, January 1). Database Systems: Design, Implementation, and Management. 13th ed. Boston, MA: Cengage Learning. ISBN-13: 978-133627900.

Moss, L. T., & Alert, S. (2003). *Business Intelligence Roadmap – The Complete Project Lifecycle for Decision Support Applications.* Reading, MA: Addison-Wesley. ISBN-13: 978-0201784206

Olszak, C. M., & Ziemba, E. (2003). Business intelligence as a key to management of an enterprise. *Proceedings of Informing Science and IT Education Conference,* 2003. Retrieved 19 February 2021 from http://proceedings.informingscience.org/IS2003Proceedings/docs/109Olsza.pdf.

Olszak, C. M., & Ziemba, E. (2004). Business intelligence systems as a new generation of decision support systems. *Proceedings PISTA 2004, International Conference on Politics and Information Systems: Technologies and Applications.* Orlando: The International Institute of Informatics and Systemics.

Olszak, C. M., & Ziemba, E. (2006). Business intelligence systems in the holistic infrastructure development supporting decision-making in organizations. *Interdisciplinary Journal of Information, Knowledge and Management*, 1, 47–58. Retrieved February 19, 2021 from http://ijikm.org/Volume1/IJIKMv1p047-058Olszak19.pdf.

Oracle Human Resource Analytics. (2020). Oracle HR analytics. https://www.oracle.com/us/solutions/ent-performancebi/hr-analytics-066536.

Perkowitz, M., & Etzioni, O. (1999). Adaptive web sites: Conceptual cluster mining. Proceedings of the 16th Joint National Conference on Artificial Intelligence. Stockholm.

Piyayodilokchai, H., Panjaburee, P., Laosinchai, P., Ketpichainarong, W., & Ruenwongsa, P. (2013). A 5E learning cycle approach–based, multimedia-supplemented instructional unit for structured query language. *Educational Technology & Society*, 16(4), 146–159.

Poul, S., Gautman, N., & Balint, R. (2003). Preparing and data mining with Microsoft SQL Server 2000 and analysis services. Addison-Wesley.

Reinschmidt, J., & Francoise, A. (2000). *Business Intelligence Certification Guide.* IBM, International Technical Support Organization. retried from http://www.sciepub.com/reference/42061

Rodeh, O., Helman, H., & Chambliss, D. (2013). IBM research report: Visualizing block IO workloads. Retrieved February 19, 2021 from http://domino.watson.ibm.com/library/cyberdig.nsf/papers/9E29FDDED06E5DE785257C1D005E7E9E/ $File/rj10514.pdf.

Russom, P. (2011). Big data analytics. TDWI Best Practices Report, 4th Quarter. Retrieved February 19, 2021 http://tdwi.org/research/2011/09/best-practices-report-q4-big-data-analytics.aspx.

SAP ERP Workforce Analytics Software. (2019). SAP business suite. https://www.sap.com/solutions/business-suite/erp/featuresfunctions/workforceanalysis/index.epxJournal.

SAS Human Capital Intelligence Software. (2020). SAS HCI. http://www.sas.com/solutions/hci/hcretention.

Singh, D., Singh, A., & Karki, S. (2021). Knowledge management and Web 3.0: Introduction to future and challenges. In *Knowledge Management and Web 3.0.* De Gruyter, Cambridge University Press, 1–14. Doi: 10.1515/9783110722789-001.

Sircar, S. (2009). Business intelligence in the business curriculum. *Communications of the Association for Information Systems*, 24(17), 289–302.

Srivastava, J., & Cooley, R. (2003). Web intelligence: Mining the Web for actionable knowledge. *Journal on Computing*, 15.

Steven Morris DATABASE SYSTEMS: DESIGN, IMPLEMENTATION, AND MANAGEMENT, 13E,Cengage Learning (1/1/2018) ISBN-13: 978-133627900

Topi, H., Valacich, J., Wright, R., Kaiser, K., Nunamaker, J., Sipior, J., & de Vreede, G. (2010). IS 2010: Curriculum guidelines for undergraduate degree programs in information systems. *Communications of the Association for Information Systems*, 26(18), 359–428.

Turban, E., Sharda, R., & Denlen, D. (2011). *Decision Support and Business Intelligence Systems* (9th ed.). Upper Saddle River, NJ: Pearson Prentice Hall.

Watson, H., & Wixom, B. (2007). The current state of business intelligence. *Computer*, 40, 96–99. Doi: 10.1109/MC.2007.331.

Williams, S. (2011). 5 barriers to BI success and how to overcome them. *Strategic Finance*, 93(1), 27–33.

Wixom, B., Ariyachandra, T., Goul, M., Gray, P., Kulkarni, U., & Phillips-Wren, G. (2011). The current state of business intelligence in academia. *Communications of the Association for Information System*, 29(16), 299–312.

Xu, L. D., Xu, E. L., & Li, L. (2018). Industry 4.0: State of the art and future trends. *International Journal of Production Research*, 56(8), 2941–2962. Doi: 10.1080/00207543.2018.1444806.

Zambon, I., Egidi, G., Rinaldi, F., & Cividino, S. (2019). Applied research towards industry 4.0: Opportunities for SMEs. *Processes*, 7(6), 344. Doi: 10.3390/pr7060344.

Chapter 5

Business Intelligence and Branding

Dileep M. Kumar
Nile University of Nigeria

Mukesh Kumar
GNS University

Normala S. Govindarajo
Xiamen University

Ipseeta Nanda
GNS University

Contents

DOI: 10.4324/9781003184928-5

5.1 Introduction

The significance of Artificial Intelligence (AI), intelligent computer programs, which has its relevance is well acknowledged, especially during the past decades. Beyond science and engineering fields, AI has proved its relevance in business sectors also imparting its significance in marketing, consumer analytics, branding and business conversions (Sanjiv, 2021). In a similar way human beings function with their intelligence, machines are programmed to operate into intelligence business platforms and works, in analysing the business scenarios and provide real-time data for managerial decision-making. Several high-profile Fortune 500 companies are engaged in applying AIs to gain advantage of the business situations, with the AI systems performing the ensuing actions with language identification, learning, scheduling and problem-solving (Chalermpol, Phayung, & Choochart, 2016). Application of automation and Information Technology (IoT) turned to be an integral part of science and engineering, and they have the very goal to create intelligent machines that support business and market. In this context, West et al. (2018) found that using AI is not a panacea to build the brand of a company. Their findings stated that using AI has a positive effect on all elements of the brand; its effects are currently the most profound on three brands: promise, customer service and personalization.

Applications of AI with Business Intelligence (BI) have a crucial role in marketing management, especially branding (Kautish, 2008; Kautish & Thapliyal, 2013). It is very fruitful to make decisions that are consumer-oriented. In marketing, it is called the consumer is the king (Bagale et al., 2021). The consumer is the only source to generate the revenue of the organization. BI is not only communicated to the potential and existing consumers, but it supports the branding of the organization. BI looks similar to technical applications, which are used for data extraction, transformation, loading and creating a report or dashboard on available business information (Magaireh, Sulaiman, & Ali, 2019). The primary goal of the BI is to assist decision maker for making better decisions. Hence, BI is a data-driven decision support system (DSS). BI is an assortment of a group of organized rules, methods, models and technologies that transform the inputted data into significant outcomes and accelerate beneficial business exercises (Luger, & Stubblefield, 1993). It is an assortment of applications and utilities to convert information into actionable insight and information. This technology has directly affected the company's strategy and business decisions. Marketers started using AI-enabled BI for effective marketing and branding.

5.1.1 Definitions of AI

A few of the earlier detentions on AI clearly stated that "it is the automation of activities which are related to human thinking, especially those activities on "arriving at decisions, explaining the problems, learning…" (Bellman, 1978). AI is the branch of study that makes use of computational models to elucidate mental capabilities (Charniak & McDermott, 1985). It is also defined by Rich and Knight (1991) that AI is to program computers to learn how to do things and thereby replace human beings, and it is concerned with automation of intelligent behaviour with the support of IoT applications (Luger & Stubblefield, 1993). Progressively, the term AI is further reviewed by new generation research scholars and is described as a field of study that integrate and analyse the computational agents that can perform an action intelligently (Poole and Mackworth, 2010). As it is specified by the researcher an agent can be anything that performs an action. The agent turned to be intelligent once

1. its activities are suitable for its situations and its goals;
2. it is bendable to varying situations and shifting goals;
3. it acquires from familiarity; and
4. it sorts right selections specifying its perceptual and computational boundaries.

While during the course of time the meaning and description of AI also has altered a system's capability to properly deduce exterior facts and figures, to acquire from such facts and figures, and to apply those learnings to attain precise goals and assignments through malleable adaptation" (Kaplan & Haenlein 2019).

5.2 AI Objectives

Several aims are associated with AI applications in general. The aims comprise Engineering, Psychological and Philosophical.

5.2.1 Engineering Aims

Extending techniques of engineering and principles of computational artefacts, which are arguably intelligent. Trying to replace human brains with automated or mechanized artefacts. Such kinds of artefacts are applicable in several domains, such as Business and Commerce, Arithmetic, Art and day-to-day life (Aron, 2021).

5.2.2 Psychological Aims

The psychological objective is to get better insight into human cognition and develop computational principles, models or structures.

5.2.3 General Philosophical Aims

To create computational doctrines, theories or structures that deliver a better understanding of cognition is common. It comprises human-made objects, naturally taking place organisms and conscious objects yet to be revealed. The philosophical objectives thus look into issues, such as the nature of intellect, thought, awareness, etc.

5.3 Subsets of AI

The most commonly used terms across the computational intelligence disciplines include acronyms and their definitions

- **Artificial Intelligence (AI)**: the all-inclusive discipline of constructing intelligent machineries
- **Machine Learning (ML)**: structures that can pick up from involvement
- **Deep Learning (DL)**: systems that absorb from involvement on bigger datasets (Sharma, 2018)
- **Artificial Neural Networks (ANN)**: prototypes of human neural networks that are intended to assist in processer learning
- **Natural Language Processing (NLP)**: systems that can comprehend linguistic aspects
- **Automated Speech Recognition (ASR)**: usage of computer hardware- and software-grounded systems to recognize and understand human speech

Engineering, science as well as disciplines linked to trade, business and commerce are making use of these terms widely to acknowledge the process integration. Nowadays, computational intelligence is widely used and seen in the domains of psychology and cognitive sciences. Computational

Intelligence models have demonstrated to be active on real-world complications in practically all disciplines, including Science, Engineering, Business, Healthcare, Management and Avionics.

The subset of AI includes three components, namely machine learning (ML), deep learning (DL) and natural language processing (NLP).

5.4 Machine Learning

ML is the subsection of AI that creates PC programs that have the right of entry to facts and figures coming from various sources by giving them the structured capability to learn and advance accordingly by identifying patterns in the database with no human mediation or activities. Based on the data type, i.e., labelled or unlabelled data, the ML model is trained to either be supervised (SL) or unsupervised learning (UL) (Jordan & Mitchell, 2015).

ML is a ground-breaking method that has widespread usage in forecasts. It is data-examining technique, which has its major concern to create computers, keep informed or adapt their activities. Such kinds of actions are in the form of predicting about any occasion or scheming an appliance, for instance, an intelligent robot (Kushwaha et al., 2020). Nevertheless, ML is always a subset of AI. ML has the capacity to learn the activities by themselves. The system performance gets improved when machines are continuously exposed to activities and perform without human intervention.

5.4.1 Categories of ML Algorithms

The ML set of rules can be mostly categorized into three groups:

- Supervised
- Unsupervised and
- Reinforcement learning.

5.4.2 Supervised Learning

An SL algorithm proceeds with an identified set of input information and its known answers to the output information, in order to study the regression or grouping model. The system will train the model to provide a prediction for the response to new information.

5.4.3 Unsupervised Learning

Only when unlabelled data is available for processing, UL will be used. The major objective behind processing UL is to infer proper patterns from the unlabelled datasets, which later provide known outputs. It is named unsupervised due to the reason where the algorithms are left on their own to group to explore the resemblances, variances and patterns in the information. When the professionals have to make use of exploratory data, UL will be applied.

5.4.4 Reinforcement Learning

Learning that is constantly intermingling with the environment is called as reinforcement learning. It is a type of ML algorithm, where an agent learns from an interrelated situation, which is mostly reflected in the form of a trial and error method, and by getting constant feedback from its former engagements and experiences. Mostly, the agents obtain rewards for acting correct ways and penalties for doing it wrongly.

5.4.5 Deep Learning

Usually, the DL is applied to heavier datasets. A considerable amount of knowledge is required in order to explain the complex and intelligent human behaviour.

AILabPage explains DL as *Irrefutably an astonishing harmonisation method used on the bases of 3 footing viz., hefty data, computing power, skills and* experience *that basically has no boundaries.*

DL is applied to forecast the unpredictable. DL makes use of a mixture of non-bio neural networks and natural intelligence rather than exploring what the agents have with them readily. DL is also called a subset of ML, which is an enormously multifarious skill set in order to attain improved outcomes from an identical dataset. Some of the advantages of making use of DL include

- **Understand the Unidentified**: In this form of Dl where the techniques develop no challenge related to the prior understanding of the domain for introspection and there is a less need for feature engineering.
- **Nothing Is Compound**: For compound difficulties, such as image, video, speech identification or NLP, DL works effectively.

5.5 Natural Language Processing

Collobert et al. (2011) defined Natural Language as a field of ML that supports computers to comprehend, manipulate and examine the information provided to them, and possibly yield human language (Haleem et al., 2020; Shetty, 2018). It is directly indicated that NLP is a sub-element of AI, and is described as "it accepts unstructured data in the form of voice, text, and image and generate a meaningful structure information" (Lilburn, 2017). In their research paper, they used the interview method to collect information from many experts. Most of the experts have given their opinion on how NLP and ML have been used to build the brand image of organizations (West, 2018). Natural Language Programming can better meet the dynamic expectations of users.

- NLP, AI and ML are now and again utilized conversely so we may get our wires crossed when attempting to distinguish between the three.
- AI is an umbrella term for machines that can reproduce human intelligence. AI is equipped with systems that have a capability to mimic human activity, such as learning from model and solving problems. NLP and ML are the sub-domains of AI.
- NLP manages how PCs comprehend and decipher human language. With NLP, machines can understand composed or spoken content and perform tasks such as interpretation, keyword extraction, segmentation, classification and more. The ML algorithms are used to deliver precise responses. Thus, ML automates these processes where the machine can automatically learn or improve itself by learning through the environment.
- Chabot's was developed by using AI algorithms, where NLP was used to decipher what clients say and what they expect to do, and ML to automatically convey more precise responses by learning and gaining from past interactions.

5.6 Concept of BI

The concept of BI is comparatively novel. For almost 40 years, the PC-enabled BI systems appeared in varied forms. Several terms are used synonymously to identify BI, namely decision support, EIS (executive information systems), and management information systems (Thomsen, 2003).

The prominence of IoT platforms and the requirement of continuous data analysis is necessitated by complex computational and diagnostic needs. In this chapter, BI systems are well-defined as system-integrated information gathering, information storage and knowledge management with the tools and techniques available to analyse and interpret multifaceted interior and competitive data to planners and decision makers. Majority of the time, the term BI is coined with a reduction of time period so that the intelligence is still valuable to make effective decision-making (Singh, 2021.

Important constituents of active BI are (Langseth & Vivatrat, 2003) as follows:

- real-time information warehousing (Chhabra, 2021),
- data mining,
- programmed irregularity and omission detection,
- proactive notifying with programmed receiver determination,
- unbroken facilitation of workflow,
- programmed learning and improvement,
- topographical data organizations,
- data visualization.

5.7 BI Applications and Tools

The tools and techniques of BI upkeep the operations and facilitate the decisions making. Major expectations from BI tools include the easy access to data related to trade and market.

Such BI tools facilitate

- EIS or DSS systems,
- Produce exports,
- Support to execute OLAP,
- Apply static data analyses and
- Data-Mining

5.7.1 Decision Support System

A computer system that supports managers in their day-to-day decision-making process is called the DSS. DSS integrates the human thought process and modelling for effective and well acknowledged decision-making.

For instance, multiple-level supply chain movement solution provides answers to logistic system (Bold, 2021).

i. **OLAP**: According to Codd and Salley (1993) OLAP or Online Analytical Processing (OLAP) is an alternative active IT solution extended for business-related decision-making. One of the advantages of OLAP is the quick calculation and efficient analysis of the trade and commerce data, which extend appropriate solutions. Consolidated commercial information are conveniently and effectively accessed by the system and provide business intelligent solutions. Applications in the form of DSS and OLAP ingeniously appear in the Business Intelligence Model.
ii. **Data Warehouses**: It is a repository where huge amounts of data are stored in a read-only form on a centralized system, where a data-mart is used to store the data in a multi-dimensional form, so the accessing of data will become more user-friendly (Tryfona et al., 1999).

5.8 Users of BI: Following Are the Users of BI

i. **The Professional Data Analyst**: The data analyst digs out the insights from the DWH using statistical tools and the BI system, which helps them to find unique bits of knowledge to outline interesting business methodologies.
ii. **The IT Clients**: The BI infrastructure is maintained by the IT users.
iii. **The Decision Makers of the Organization**: The proprietor of an organization can improves the gain of their business by creating operational execution by utilizing BI in their business.
iv. **The Business Clients**: The prominent business users can access the data warehouse while the other users can access the dashboard, Map, KPI, which is developed over the historical stored data.

5.9 How BI Systems Are Implemented

i. Raw data are extracted from the different data sources of the corporate, which are stored in heterogeneous systems.
ii. The data are modified and cleaned before being stored in the data warehouse.
iii. Using such a system, the clients can raise questions and request for reports, or perform different types of analysis.

5.10 Benefits of BI

i. **Increase Productivity**: BI tools can create a complex report in a single click so employees will become more efficient and productive.
ii. **To Increase Visibility**: BI further supports with refining the visibility of these processes and makes it plausible to recognize any areas that need consideration.
iii. **Fix Accountability**: The user of the BI system will become more accountable because they assist the decision maker for making their decision.
iv. **It Gives a Higher View**: The decision maker gets a higher view of their data by using the BI tools, such as dashboards and scorecards, KPI, Slicer, etc.
v. **It Smoothens Out Business Processes**: By using BI predictive analysis, computer modelling, bench-marking and other methodologies, the analysis will become automatic.
vi. **It Allows for Simple Analytic**: BI application interface is so user-friendly due to which non-technical clients can easily access their information. So, its popularity will increase.

5.11 Role of BI in Marketing

i. **Focused Demographics and Improved Audience Sketching**

When the marketers get vague data from the field, a majority of the advertising and promotion campaigns succumb to failure. Markets usually spend heavy amount for these campaigns and there is less return of investment. It is expected that the marketing campaigns should reach to right listeners and viewers at the right time using the right channels. In this context, BI turned to be a solution, the power to efficiently build strategies, by gathering and analysing all information related to demographic diversity, such as listeners and viewers, pain points, communication patterns, buying habits, purchasing capacity, etc.

ii. **Optimized Marketing Campaigns**

Tracking the customer journey and the market is the success formula for any business. Marketers may be spending large amounts of money for marketing campaigns. Making use of a BI tool will track and examine the performance of marketing and sales promotion campaigns in real-time and equate it alongside historical patterns. Frequent contrast and comparison can support the marketers in making an effective decision on how to re-arrange the marketing accounts and course it to advertising activities. Such applications support companies to get the highest return on investment. In addition, with the market data coming from varied sources, BI tools support the markets to execute up-sell and cross-sell campaigns. BI tools enable to take proper decisions on price points, market overstocked items and slow-moving merchandise with better visibility of customer purchasing habits. With BI tools, and applications, firms will be endowed with a choice to augment customer acquirement and ease the prospects towards sales.

iii. **Speedier Reporting Process and Quality Insights**

Understanding targeted market is an important aspect for every marketer to fine tune marketing decision with the market volatility. A dashboard with data coming from all sources is important for decision-making. A dash board with data visualization will extend full coverage of the market with charts and graphs, suggesting the firms to have corrective or responsive strategies for better market position. The report process will be speedier enough to save time and facilitate better computational decision-making. Companywide issues related to customer contentment and developing strategies for market expansion is the crux of BI tools application.

5.12 New Marketing Strategies and Digital Marketing

Digital marketing with the support of BI tools enables the managers to float the product and service information through several publicizing means, such as search engines, email, blogs, social network and product websites. These medium of communication generates enormous amount of data, which will facilitate analysis and recycling of established marketing structure and its lucidity is well ascertained. The features of assistive technology support many firms to have data in their fingertips that ascertain their rapid growth. As it is evidenced the most prevalent use of technology in advertising and promotion is social media. The easiness of its use among marketers has created an image of perceived usefulness in business growth. Some new marketing strategies empowered with BI tools include the following:

5.12.1 Content Marketing

BI-enabled digitalized content marketing and several brands make use of it through social media, multimedia and mobile search. BI-supported digitalized content marketing is very imperative to build brand awareness.

5.12.2 Mobile Marketing

Excessive use of smartphones enables BI-supported digitalized marketing for several brands, which has proved effective when customers can see those campaigns any time ensuring flexible accessibility. BI-enabled apps facilitate the companies to redesign their websites to make them reactive to mobile devices.

5.12.3 Integrated Digital Marketing

BI-enabled digitalized combined marketing communication is significant to guarantee that all messaging and customer interactive strategies are integrated across and circled on the customer. For example, Google has invented Google+to view and gather social signals and patterns.

5.12.4 Continuous Marketing

Digitalized synchronized BI tools constantly remind the consumers about the product and services. This tactics is one of the widespread marketing strategies that create brand awareness.

5.12.5 Personalized Marketing

IoT-synchronized BI tools support marketers to read and craft personalized messages. BI tools will read consumer's behaviour and extend them with relevant information for effective marketing and promotion decisions.

5.12.6 Visual Marketing

Instagram, slideshare, Pintrest and several platforms are BI-enabled visual marketing strategies to put advertising images, messages and signals. BI-enabled digital marketing strategies are the trend for many brands.

5.12.7 Search Engine Optimization

BI-enabled search engine optimization (SEO) is the method of enhancing the number of visitors to a specific website by confirming that the site seems great in the list of outcomes refunded by a search engine. The SEO empowers companies to produce a continuous stream of organic traffic to their websites. SEOs are a basic ingredient of digital marketing efforts of the companies since the consumers are highly engaged in search engine options to identify the products and services in accordance with the taste and preferences.

5.13 BI and Consumer Analytics

Sentiment analysis comprises methods, techniques, and tools for extracting subjective information, such as opinions and attitudes from language; the object is the product or service about which the reviews are shared by the customer, the emotionally loaded opinions. Sentiments are feelings or to some degree of emotional importance and are typically personal. Sentiment analysis describes an assemblage of methods that attend the problem of gauging view points, feelings and prejudices in texts. However, the study of bias, prejudices, emotion, etc., are within the domain of numerous subjects, including literature, history, political science, etc.

Sentiment analysis is usually a branch inside computational dialectology; however, it is commonly considered as a method in social science. Consumer delight is one of the important objectives of marketing management. Most of the micro and macro factors are influencing the consumers' personality and consumers start purchasing accordingly. Subject well-being is directly associated with customer fulfilment and examines the importance of subjective well-being in consumer sentiment. The research found that subjective well-being was measured by the International Well-being Index and contributed significantly to the description of consumer sentiment and made a situation for its addition in more customer satisfaction (Ganglmair & Lawson, 2012).

BI-enabled customer analytics permits users to respond to multifaceted questions in a more collaborating manner. Customer analytics applications and tools integrate diverse types of occasions and activities to construct a complete representation of a customer journey. This indicates that the marketers can observe the exact paths that a user commences from one location to the next, to achieve an exact objective. BI-enabled customer analytics thus support effective understanding of people and they can work together in more cooperative and creative ways—nurturing a culture of coordination and innovation.

BI has continuously been about generating quicker, healthier and more precise procedures for turning disjointed understandings into actionable stories. For instance, with the introduction of Amazon Web Services (AWS) in 2006, the computing supremacy touched an unparalleled level of scalability. As a result, SaaS BI platforms have turned out to be so effective, reachable and cost-effective for any type of company, and have the potential to become a data-driven operation. One of the astonishing market trends currently is the integration of AI into all BI operations enhancing the value propositions. Competitive footing of any firm in the market, around the word, needs data capabilities and BI support in the data integration for decision-making.

5.14 Advantages in Adapting AI-Enabled BI Systems

A. To customers:
   i. AI-enabled BI systems can support consumers 24×7.
   ii. AI-empowered BI applications effortlessly trace consumer behaviour and predict upcoming Web behaviour and decisions.
   iii. AI-facilitated BI systems preserve customer statistics and information and there is no need to repeat with every interface.
   iv. AI-assisted BI systems are user-friendly and at all times delight customers graciously and with tolerance.
   v. AI-empowered BI applications can hold numerous consumer's demands concurrently, which reduces the waiting time.

B. To Marketers:
   i. AI-enabled BI systems are equipped with big data insights, where the digital marketers can boost marketing campaigns, which facilitate return on investment.
   ii. There is only little physical work and little errors.
   iii. BI systems make sure that the correct message is being provided to the right individual at the right spell, using the network of select.
   iv. AI-supported BI systems promote brand image of the firm and shape more potential consumers, which leads to better sale of products and services.
   v. BI-enabled customer-tailored advertisements can be generated for enhancing sales.
   vi. AI-enabled BI systems have the power to track consumer buying patterns, which permits the corporate to renovate current advertising plan and increase sales.
   vii. BI systems perform as effective way to produce an association with the customers that has high significance.

5.15 AI, Technology Mapping and Branding

i. **Necessity of Adapting AI for Business Intelligence: As a Marketing Strategy**

Active sales and promotion plans will yield accurate decisions, which enable firms to be successful whether they are IOT based or offline. To attain a positive promotion strategy, it is essential to follow up on new social media trends that maintain constant interaction with consumers. Accordingly, the power of AI-supported BI structures in digitally enabled promotion plans permit a salesperson to promote his/her products or services and succeed in trade processes. AI applications are the foremost topic widely discussed by leaders across organizations. AI covers a broad range of technology-driven applications that try to program human intelligence, integrating human competences, namely voice and image identification, ML systems, semantic searches, voice recognition and averting data leaks ensuring smooth accomplishment of task-driven processes. Marketers are even thinking about making use of drone applications at remote sites. One of the advantages is that all kinds, forms and sizes of businesses can integrate AI for any functional activity linked to marketing and branding.

ii. **Illustration: AI in Fashion**

Application of AI is rapidly progressing in major retail operations as well as product developments. Several big brands, such as Alibaba, have introduced their leading "Fashion AI" idea, almost dated back to 2018. Alibaba has given marvellous Artificial Intelligence Experience to customers with the integration of RFID racks and AI mirrors. Such AI applications have enhanced consumer experience to a whole new level.

Similarly, one of the giant retailers such as Tommy Hilfiger too has given customer experience with the integration of AI in fashion. IBM and the Fashion Institute of Technology (FIT) have jointly created a new tagline viz., `Reimagine Retail to get traders a better position in terms of quickness that permits designers to craft good quality, more personalized pieces (Carla, 2021).

The BI system architecture comprises a set of functions and tools that help to meet business goals. The applications of BI are designed with a more user-friendly interface. The BI application has the ability to generate standard and real-time simple and complex reports. The BI application development follows data validation, testing, maintenance and improvement phase (Kautish & Thapliyal, 2013).

iii. **Why is BI Important?**

a. Such frameworks distinguish market patterns and identify business issues that ought to be tended to.
b. Such frameworks assist with data report in the form of slicer, dashboard, etc., which improves the information quality and subsequently the standard of decision-making.
c. Such frameworks can be utilized by undertakings as well as Small and Medium Enterprises.

Integrating AI applications and tactics into customer lifecycle, the below-mentioned diagram extends better understanding on computer intelligence applicability across the customer journey.

5.16 AI, BI and Brand Strategy

A selection of shared and idiosyncratic brands features a firm spread across the various products and services it vends and the firm itself is called as its brand strategy. It replicates the number and kind of fresh and current brand features, although giving proper direction on how to brand novel products. A brand strategy comprises systematic structuring and layout of an imminent image for the firm, extending a course of action and standards against which to evaluate it. A brand strategy will be formed based on future goal line. One of the prominent aspects related to brand strategy is disseminating continuous attentiveness about the products to increase brand awareness, generate a constructive brand image and find brand inclinations and brand loyalty. The brand strategy also targets at developing the charm and desirability of the firm in the eyes of the stakeholders who support the administration of the firm, and to provide the staffs with the standards and the values of their own engagements. AI- and BI-applied predisposition models, which can be applied at this phase to draw attention among more invitees and deliver those who hit your company websites, provide a more appealing experience.

5.16.1 AI-Generated Content

Attracting visitors to the website needs appealing content to the website. Although AI has limitations in writing website content based on the changing preferences of the customers, AI content writing courses are able to select components from a dataset and construct a 'human sounding' piece of information. For instance, an AI writing program called 'WordSmith' created 1.5 billion bits of content in 2016, and is predicted to grow further in the future. AI writers contribute highly for detailing contents, which are based on daily events. Dynamic information's can be better plotted in the website through these data-focused AI writers, for instance, monthly earnings reports, sports matches and market information. Especially on some business niche areas, such as financial services, AI-generated content would be a highly obliging component of your content marketing strategy (Allen, 2017).

5.16.2 Smart Content Cure

Engaging the visitors with appropriate content in the website is one of the major tasks, which can be assigned to AIs. AI-driven content cure lets one to engage effectively on the website by displaying the content pertinent to them. Some of such content can be significantly seen in the websites, such as 'consumers who purchased X also obtained Y' segment on several sites, but can also be functional to blog content and individualizing website messaging more extensively. The AI-driven content curation is also a prodigious tool for subscription businesses, where someone uses the ML algorithm to apply and improve the endorsements of content. The best example in this context is Netflix's suggestion system. Netflix regularly suggests the shows one could be interested in (Allen, 2017).

5.16.3 Voice Search

Voice exploration is an additional AI tool. Nevertheless, when it relates to applying for promotion and sales, it will lead to some of software giants who are the major players in the market, such as Google, Amazon, Apple. Voice search is going to redefine the prospect of SEO strategies, and brands required to keep up. In organic traffic, any brand that makes use of voice search can better take advantage of the high buyer intent due to AI-led computer-generated personal assistants (Allen, 2017).

5.16.4 Programmatic Media Buying

Programmatic media purchases may make use of predisposition models generated by ML algorithms to competently aim advertisements at the most important consumers. Programmatic advertisements require to get keener in the wake of Google's latest brand safety disgrace. It is indicated that the advertisements placed programmatically through Google's ad network were sighted on terrorist's websites. Doubtful and suspicious websites can be identified using AI tools and techniques and removed from the list of site advertisements they are positioned on.

5.16.5 Propensity Modelling

Propensity modelling is the main purpose of an ML task. The ML algorithm is served with huge volumes of past data, and it applies this information to produce a propensity model that (in theory) can accurately predict about the real world. (Allen, 2017).

5.16.6 Predictive Analytics

Propensity modelling may be applicable to several distinguished areas, such as predicting the intention of a consumer to choose 'a price', with which the customers can be converted into for 'real buying' and make 'repeated buying'. This application is called predictive analytics, since it makes use of analytics data to make forecasts about how consumers behave. Here mistakes may happen if the random data obtained from the site are incurred with errors. Accurate predictions are possible only when the data are reliable and the measurements are valid (Allen, 2017).

5.16.7 Lead Scoring

Propensity models created by ML may be well accomplished to score leads based on selective criteria intended by the firm, so that the sales team can find how significant a given lead is and whether they are worth noting to give ample time for decision-making. The lead generated is very much significant in the case of B2B business, where there are consultative sales processes. B2B sales process is tough and consumes a lot of time. Identifying and approaching is a most significant and important aspect for business conversion. AI tools can support firms to get proper insights and supply those leads for targeted sales and discounts.

5.16.8 Ad Targeting

A vast amount of historical data can be run with the support of ML algorithms. The advertisement can make a best prediction on the type of people and stage with which buying process will happen with ML algorithms. Continuous optimization of the data can be made possible through ML algorithms, and thereby firms can achieve more effective ad placement for business conversion (Allen, 2017).

5.16.9 Dynamic Pricing

Shifting into more products will produce better sales and almost all marketers are well aware of this aspect. Discounts and rebates are effective with almost all segments of customers, although it has its impact at the bottom line. ML can develop a propensity model in which predictions can be made on people who will be ready for a purchase with an offer. Without reducing the profit, and dynamic pricing, one can fix the sales proposition with the support of ML algorithms (Allen, 2017).

5.16.10 Web and App Personalization

Making use of a propensity model journey of the customers can be predicted well and where the conversion of sale will be happening can be determined. The Web content can be corrected and made attractive enough with relevant content for better sales conversion. Contents in the websites should be more effectively accommodating frequent visitors and those who are moderately using the websites (Allen, 2017).

5.16.11 Chatbots

Chabots are tools that mimic and copy human intelligence and support human enquiries in the digital platform. Chatbots engage in almost 100% sales interactions and support in sales conversions. For instance, Facebook desires to make its Messenger app the go-to place for individuals to engage in chats with brand's virtual ambassadors. This indicates that brands may make use of Facebook's powerful bot development tools (Allen, 2017).

5.16.12 Re-targeting

Bringing back customers to brand is a tedious task since they have already made up their mind to switch over to other ones. One of the advantages of AI-based business intelligent applications is that ML can be applied to create what kind of website contents can be integrated into so that the firm can fetch consumers back to the site, which is based on past data. ML can be applied to enhance your re-targeting advertisements to make them effective (Allen, 2017).

5.16.13 Predictive Customer Service

Attracting new customers will be a herculean task than maintaining existing customers. A business that is based on subscription bases has to face heavier challenges and the churn rate can be extremely costly. Those customers who are most likely to unsubscribe from a service can be brought back with the support of predictive analytics. With the support of ML algorithms, such customers can be identified, and potential to reach out with offers, reminds or supports to prevent them from churning (Allen, 2017).

5.16.14 Marketing Automation

One of the good features related to marketing automation is the continuous interactions with consumers (Eshna, 2021). A marketer has to decide the most effective way of algorithms that has to be integrated with the website. Algorithms of ML can operate through billions of points of consumer data. It will support the most effective timing to contact, the words to be used and the way of interaction. Marketing automation insights can enhance the effectiveness of market communication and improve the conversion of business (Allen, 2017).

5.16.15 Dynamic Emails

Predictive analytics, which makes us a propensity model, could be able to buy certain categories, sizes and colours by analysing through customers their previous behaviour and displays the furthermost pertinent products in bulletins. All information regarding the product, its stocks, deals and pricing details need to be accurate when they are updated on a website. Businesses can successfully administer ML projects and make use of the insights generated to improve marketing results.

5.17 BI and Branding Short Case Studies

A. **Case 1: Starbucks Uses Predictive Analytics to Serve Personalized Suggestions**

Case studies directed that research firms that recognize customer requirements and desires through predictive analytics are capable enough to raise their income by 21% year-over-year in comparison with an average of 12% deprived of predictive analytics (Dan, 2021).

The case study cites one of the major examples from Starbucks, which makes use of its loyalty card and mobile app to gather and analyse data. Starbucks uses predictive analytics platform to process orders coming through apps and serves consumers with tailor-made promotion messages. These communications comprise suggestions and endorsements when customers are approaching a local store as well as offers that have the objective to enhance the average order value. AI-based BI applications supported Starbucks in brand management. Through improved brand management methods, customer loyalty is attained. Such decisions have turned around the Starbucks with better revenue, especially in terms of improved sales. Starbucks embraced BI applications that emphasizes on enlightening their brand management. Starbucks could improve customer experience with the application of AI-based BI applications. The satisfied customers are the brand ambassadors. AI-based interactivity supported in spreading the word about Starbucks to other customers. Another approach is to use AI to monitor Starbucks brand on social media and other digital media. It has enhanced the reputation of the brand across the product line. Effective marketing engagement has thus enhanced brand awareness. AI-enabled BI applications thus have huge possibilities in delivering advanced brand management.

B. **Case 2: Alibaba Opens a Fashionai Store**

With the objective of extending augmented retail experience to consumers, Alibaba opened a physical store, namely 'FashionAI', in Hong Kong. Augmented features with AI and BI applications include intelligent garment tags. When the product is touched the smart mirrors demonstrate apparel information and recommend synchronizing products. Alibaba has also revealed its upcoming brick-and-mortar store with a virtual wardrobe app. This app will permit consumers to understand the garments they tried on in-store. The decision to adopt and transform into technology-based apps is rooted with the shifting expectations of consumers. A recent study conducted by Retail Dive (2017) clearly reported that there is an increase of 46% of those surveyed believed to have positive experiences with AI and BI and has given them more assurance in a particular brand.

C. **Case 3:** Customers Create Custom Nikes **in 90 Minutes**

In order to allow consumers to design their own sneakers in store, in 2017, Nike floated a system 'Nike Makers' experience'. This system allows consumers to put on blank Nike Presto X sneakers and select their own visuals and colours. Making use of augmented reality and projection schemes, the system then exhibits the design on the blank shoes. The designs are reproduced on the sneakers and obtainable to the consumers within 90 minutes. Such a rare customer engagement experience can enhance their impressions of the brand and increase the sales. Nike used AI-based ML algorithms to collect the data and design forthcoming products and deliver individualized product suggestions and promotional messages.

D. **Case 4: Amazon Launches Personalize**

Amazon was a forerunner of applying ML to deliver individualized product suggestions. Amazon broadcast the overall accessibility of Amazon Personalize that fetches the similar ML technology applied by Amazon.com to AWS consumers in order to use it in their applications (AWS-CTO, 2018).

Brands comprising Domino's, Yamaha and Subway are by this time using *Personalize* to high spot musical instruments in-store collections, supply ingredients and flavour suggestions, and devise personalized style groupings.

5.18 Risks and Limitations in AI-Enabled BI in Marketing and Branding

i. AI-enabled BI is limited only by the obtainability of information from varied sources.
ii. Frequent assignments can easily be removed over to AI-enabled BI applications, but assignments need human intervention for effective operations.

iii. AI-enabled BI systems incur significant amount of cost of acquisition and maintenance. The ROI needs to be sensibly reflected before its implementation.
iv. Application of AI-enabled BI systems is time-consuming.
v. Algorithms of AI-enabled BI systems can be wrong as PCs cannot do the operations without humans.
vi. Consumer privacy is not taken into consideration.

5.19 BI System Disadvantages

i. **Cost**: The BI system is very expensive for medium- or small-sized organizations. For routine business transactions, the use of such a system is very expensive.
ii. **Complexity**: Its implementation is very complex.
iii. Limited use of the cost of the BI technologies was very expensive so only the large capital size organization could purchase the software.
iv. **Time-Consuming Implementation**: The internal architecture of the data warehouse is very complex so its implementation into an organization requires more time.

5.20 Conclusion

How a firm approaches its decision-making is highly influenced by the application of AI and BI, by making use of information obtained from varied sources. Organizing firm's goals using AI and BI turned to be a crucial aspect. AI enables computer systems to contemplate intelligently like human beings. BI provides facts and figures into significant insights in the form of reports and dashboards – which enables firms to come up with better data-driven decisions. Brand management for marketers turned easier with the support of AI and BI. Google and Microsoft are among the foremost global brands that have calibrated their focus on AI applications for better brand management. Other firm's leaders, such as Alibaba, TenCent, Amazon, Facebook and Apple, are also following the pathways of AI and BI. This indicates that AI and BI are fetching the core know-how for augmenting consumer engagement and managing brands.

References

Allen, R. (2017). *Mapping the Most Effective AI Technologies for Marketing across the Customer Lifecycle.* Cited in: https://www.linkedin.com/pulse/15-applications-artificial-intelligence-marketing-robert-allen/ Accessed on 28 Feb 2021.

Aron. (2021). *How Artificial Intelligence is Fashion's Perfect Fit.* Online Article: https://www.chainofdemand.co/how-artificial-intelligence-is-fashions-perfect-fit/ (Accessed on 12 June 2021).

AWS-CTO. (2018, November 29). Amazon launches personalize, a fully managed AI-powered recommendation service. https://venturebeat.com/2019/06/10/amazon-launches-personalize-a-fully-managed-ai-powered-recommendation-service/ (Accessed on 10 May 2021).

Bagale, G. S., Vandadi, V. R., Singh, D. et al. (2021). Small and medium-sized enterprises' contribution in digital technology. *Annals of Operations Research.* Doi: 10.1007/s10479-021-04235-5 (Accessed on 17 July 2021).

Bellman, R. (1978). *An Introduction to Artificial Intelligence: Can Computers Think?* San Francisco, CA: Boyd & Fraser Pub. Co.

Bold, B. (2021). *Solutions for Supply Chain Management.* Online Article: https://www.boldbi.com/dashboard-examples/supply-chain (Accessed on 18 July 2021).

Carla, G. (2021). *Applications of Artificial Intelligence in Marketing.* Online Document: https://www.pinterest.co.uk/pin/281334307955598814/ (Accessed on 21 August 2021).

Charniak, E., & McDermott, D. (1985). *Introduction to Artificial Intelligence.* Reading, MA: Addison-Wesley.

Chalermpol, T., Phayung, M., & Choochart, H. (2016). TLS-ART: Thai language segmentation by automatic ranking trie. *Conference Paper. The 9th International Conference Autonomous Systems (AutoSys 2016)*, Cala Millor, Spain. https://www.researchgate.net/publication/311705165_TLS-ART_Thai_Language_Segmentation_by_Automatic_Ranking_Trie (Accessed on 21 August 2021).

Chhabra, M. (2021). *Data Warehousing and Business Intelligence Simplified 101, on BI Tool, Data Integration, Data Warehouse, ETL.* HEVO. Online Document: https://hevodata.com/learn/data-warehousing-and-business-intelligence/ (Accessed on 20 August 2021).

Codd, E. F., Codd, S. B., & Salley, C. T. (1993). *Providing OLAP (On-Line Analytical Processing) to User-Analysts: An IT Mandate, E. F. Codd and Associates.* White paper (sponsored by Arbor Software Corporation).

Collobert, R., Weston, R., Bottou, L., Karlen, M., Kavukcuoglu, K., & Kuksa, P. (2011). Natural language processing (almost) from scratch. *The Journal of Machine Learning Research*, 12, 2493–2537.

Dan, R. (2021). *How STARBUCKS is Using Artificial Intelligence to Connect with Customers and Boost Sales.* Online Document: https://finance.yahoo.com/news/starbucks-using-artificial-intelligence-connect-232830287.html (Accessed on 21 July 2021).

Eshna, V. (2021). *How AI and Automation Are Changing the Nature of Work*. Online Article: https://www.simplilearn.com/how-ai-and-automation-are-changing-the-nature-of-work-article (Accessed on 21 June 2021).

Ganglmair, W., & Lawson, R. (2012). Subjective wellbeing and its influence on consumer sentiment towards marketing: A New Zealand example. *Journal of Happiness Studies*, 13(1), 149–166.

Haleem, A., Javaid, M., & Vaishya, R. (2020). Effects of COVID 19 pandemic in daily life. *Current Medical Research and Practice*, 10(2), 78–79. Doi: 10.1016/j.cmrp.2020.03.011.

Jordan, M. I., & Mitchell, T. M. (2015). Machine learning: Trends, perspectives, and prospects. *Science* 349, 255–260. Doi: 10.1126/science.aaa8415.

Kaplan, A., & Haenlein, M. (2019). Siri, Siri, in my hand: Who's the fairest in the land? On the interpretations, illustrations, and implications of artificial intelligence. *Business Horizons*, 62, 15–25.

Kautish, S. (2008). Online banking: A paradigm shift. *E-Business, ICFAI Publication, Hyderabad*, 9(10), 54–59.

Kautish, S., Singh, D., Polkowski, Z., Mayura, A., & Jeyanthi, M. *Knowledge Management and Web 3.0: Next Generation Business Models*. De Gruyter, Berlin.

Kautish, S., & Thapliyal, M.P. (2013). Design of new architecture for model management systems using knowledge sharing concept. *International Journal of Computer Applications*, 62(11).

Kushwaha, S., Bahl, S., Bagha, A. K., Parmar, K. S., Javaid, M., Haleem, A., & Singh, R. P. (2020). Significant applications of machine learning for COVID-19 pandemic. *Journal of Industrial Integration and Management*, 05(04), 453–479. Doi: 10.1142/s2424862220500268.

Langseth, J., & Vivatrat, N. (2003). Why proactive business intelligence is a hallmark of the real-time enterprise: Outward bound. *Intelligent Enterprise*, 5(18), 34–41.

Lilburn, A. (2017). *Impact of AI on Brand* [Interview] 2017. Cited in. West et al., (2018). "Alexa, build me a brand" An Investigation into the impact of Artificial Intelligence on Branding, *The Business and Management Review*, 9(3), 321–330.

Luger, G. F., & Stubblefield, W.A. (1993). *Artificial Intelligence: Structures and Strategies for Complex Problem Solving*. The Benjamin/Cummings.

Magaireh, A. I., Sulaiman, H., & Ali, N. (2019). Identifying the most critical factors to business intelligence implementation success in the public sector organizations. *Journal of Social Science Research*, 14, 2395–2414 Doi: 10.24297/jssr.v14i0.8026 (Accessed on 13 June 2021).

Nike Will Let You Make Your Own Shoes in Under 90 Minutes. The next generation of custom sneakers. Online Document: https://solecollector.com/news/2017/09/nike-makers-experience-custom-sneakers-presto (Accessed on 20 August 2021).

Poole, D., & Mackworth, A. (2010). *Artificial Intelligence – Foundations of Computational intelligence*, 2nd Edition, Cambridge University Press.

Rich, E., & Knight, K. (1991). *Artificial Intelligence.* (2nd edition), McGraw-Hill, New York.

Sanjiv, L. (2021). *Applications of Artificial Intelligence in Marketing.* Online Article: https://medium.com/@sanjivla31/15-applications-of-artificial-intelligence-in-marketing-eda546c402e7 (Accessed on 20 June 2021).

Sharma, V. (2018). *Deep Learning (DL) – Introduction to Basics.* Available: https://vinodsblog.com/2018/11/05/deep-learning-everything-you-need-to-know/ (Accessed on 20th June 2021).

Shetty, S. (2018). *Why Tensor Flow Always Tops Machine Learning and Artificial Intelligence Tool Survey PA.* Packtpub. Available: https://hub.packtpub.com/tensorflow-always-tops-machine-learning-artificialintelligence-tool-surveys/. Accessed on March 11, 2020.

Singh, D. (2021). Knowledge management and Web 3.0: Introduction to future and challenges. In *Knowledge Management and Web 3.0: Next Generation Business Models*, De Gruyter, 1–14. Doi: 10.1515/9783110722789-001.

Thomsen, E. (2003). BI's promised land. *Intelligent Enterprise*, 6(4), 21–25.

Tryfona, N., Busborg, F., & Borch Christiansen, J. G. (1999). starER: A conceptual model for data warehouse design. In *Proceedings of the ACM International Workshop on Data Warehousing and OLAP*, Kansas City, Kansas, 3–8.

West, M. D., Labat, I., Sternberg, H., et al., (2018). Use of deep neural network ensembles to identify embryonic-fetal transition markers: Repression of in embryonic and cancer cells. *Oncotarget*, 9, 7796–7811.

Chapter 6

Role of Business Intelligence and HR Planning in Modern Industrialization

Nitin Aggrawal
Department of Public Enterprises

Adith Potadar
Max Kelsen

Contents

DOI: 10.4324/9781003184928-6

6.1 Business Intelligence

The technological advancements, such as the Internet of Things, AI-enabled predictive modeling, robotics, Big Data-driven dashboards, process automation, intelligent process modeling, etc., have transformed the present manufacturing processes. Early adopters of such tools and technologies are now able to deliver better quality products with increased output, reduced wastage in relation to both material and human efforts. However, the successful adoption of technologies in an organization depends largely upon the agility of its human resource (HR). Thus, an organization requires more informed decision-making capabilities in terms of workforce that is technically talented and suited to the role, and retaining them by providing adequate benefits, work–life balance and through real-time monitoring of their engagement, performance, training, and other safety needs. The present HR department operates through divisional levels where most of the monitoring of the employees is delegated to divisional heads (Singh & Gite, 2015). However, micro-level monitoring of HR functions can be undertaken through technology to give business insight through data analysis and intelligence capability.

As per the ADP Research Institute engagement survey of Indian workers (Narayan, 2019), more than three-fourths of workers were not contributing their best. The reasons for such a gap were attributed to individual stress, workplace stress, low priority toward engagement by leadership, and a lack of effective training for employees. Thus, the development of Business Intelligence (BI) tools is of paramount importance and an effective mechanism that indirectly ascribes to increase employee engagement, higher productivity, and realization of business goals.

The term Business Intelligence was first coined by H.P. Luhn in 1958 (Tutunea et al., 2012). It is a combination of people skills, technologies, software applications, and business processes (Kapoor, 2010). Thus, BI is driven by technical processes to analyze data and deliver actionable information for all executives of the organization for better decision-making. This technical process is well described by the Data Information Knowledge Wisdom (DIKW) model (Rowley, 2007) (Figure 6.1). One cannot draw meaning from the data collected either internally or from external sources. These data are merely a collection of individual facts, figures, or measurements. To extrude "Information" or provide context to this data, it must be processed by ways of organizing, structuring, categorizing filtering or condensing. This information can now be represented using BI tools and dashboards to provide meaning, provide new learning, induce discussions, deliberate on notions leading "Knowledge." This deliberation or acquired Knowledge leads to "Wisdom" with the input of continued insight, which can be actioned

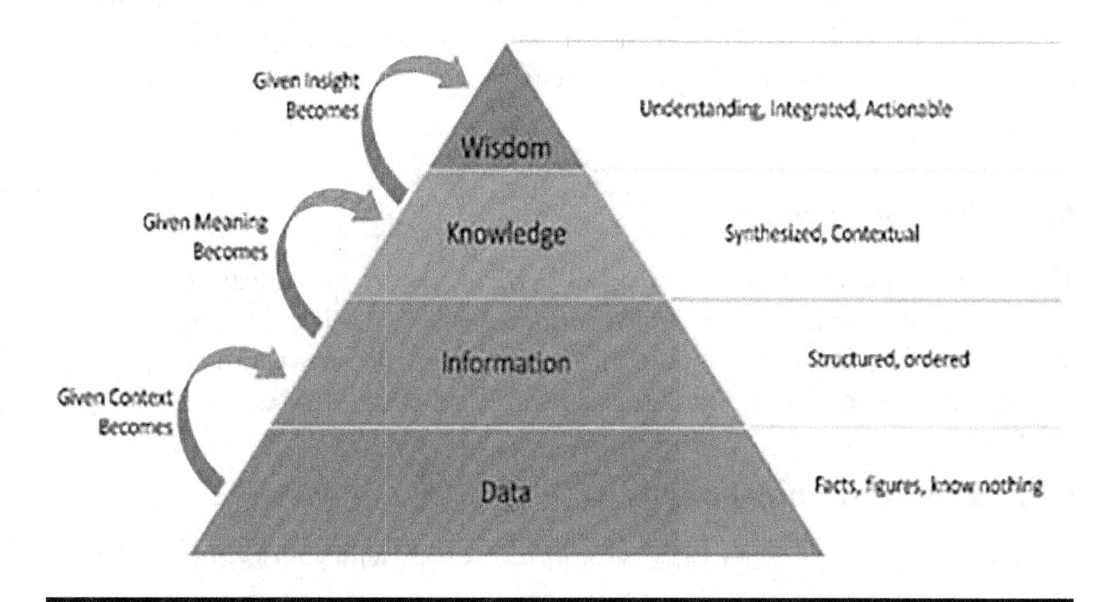

Figure 6.1 Authors' compilation. (Adapted from DIKW Model Mutongi, 2016.)

upon, reflected upon, and ultimately lead to the heightened strategic decision-making process.

BI tools are software applications that collect, process, analyze, and visualize large volumes of data. These tools create interactive reports to give a simplified view for more informed decision-making. These tools have unique features for data handling, such as Data pipelining, for real-time or near real-time updates, data cleansing, data transformation and normalization, visualization, analytical insight reports, dashboards, and scorecards (Kautish, Singh, Polkowski, Mayura & Jeyanthi, 2022). They are mostly automated utilities that can be used in self-service mode for predictive analytics. These utilities can operate on enormous amount and increasing diverse set of data coming in from various sources. They bring real-time BI capabilities to enterprises.

6.2 Significance of BI

Business intelligence (BI) empowers organizations to access information that are critical to the success of one or more of its operations, such as sales, production, logistics, finance, marketing, operations, etc. (Kautish, 2008; Kautish & Thapliyal, 2013). The operational data are captured on identified indicators. Subsequently, data are analyzed, and BI algorithm forecasts actionable insights on industry trends (Bagale et al., 2021). The following are some of the important reasons to adopt BI in business operations:

6.2.1 Business Expansion

The organizations that are into market expansion can take advantage of BI to get customer trends for their products vis-à-vis their competitors. Moreover, this descriptive trend can be used to create new products, product improvements, or identify niche markets and get desired sales output.

6.2.2 Process Improvements

BI provides better control of the operations as it provides cross-functional insights that are relevant in identifying areas for process improvement and motivating proactive actions.

6.2.3 Predictive Actions

Although BI itself is just a comprehensive description of present and past events, but when it is used with business analytics tools, it becomes a great tool for making predictive actions for the business.

6.2.4 Productivity Improvements

With a BI System, the organization is able to capture, store, and share vital information that is essential for quick decision-making. This also reduces dependency, duplication, and inconsistency. As a result, the response time and decision-making improve across all sections of the organization leading to increased throughput with better quality and lowered cost.

6.2.5 Customer Retention

Invariably, the biggest challenge for almost all organizations is to retain old customers and convert customers to competitors. The organization needs to build a layer of BI systems over its Customer Relationship Management (CRM) applications. This combination shall guide the organization on identification of critical information and its structure that can be easily converted into strategic initiatives to potential customers and retaining them through a focused services approach.

6.2.6 Real-Time Data

The present system of decision-making is cumbersome, where the department collects data pertaining to specific issues only at specific intervals. The data are analyzed and culminated into a Management Information System or a Decision Support System report for the management. The process is primarily manual and prone to human errors and is always at risk to become outdated at the next interval of data collection. However, if a systemic BI tool is used, the data get collected on real-time basis and their analytical reports and recommendation areas are always available for decision-making through visual dashboards.

6.2.7 Competitive Advantage

Besides facilitating business decision-making, BI tools can also provide insight of competitors in terms of their operational cost, logistics, and financial capability. Thus, BI can be used to study the strengths and weaknesses of competitors and draw up future plans to compete with them.

6.2.8 *Quality*

BI systems in quality monitoring have and will continue to excel in any industry. They can be applied for continuous monitoring of production process, defect analysis, quality compliance, and can act as an enabler for any kind of data-driven quality monitoring and research undertakings (Kalchauer, Lang, Peischl, & Torrents, 2014).

6.3 Purpose and Benefits of BI

The aim of BI is to provide HR managers with better insight into their HR strategies. It assists leaders in better decision-making to create an environment that fosters employee engagement and productivity. Recently, the concept of HR dashboard has become a new trend in modern industries. The HR dashboard is a BI tool that allows organization to keep track, analyze, and report on HR Key Performance Indicators (KPIs). Companies also use BI to cut costs, identify new business opportunities, and spot inefficient business processes. As per survey (BARC, 2021), some of the potential benefits identified after the use of BI tools are like increase in efficiency and accuracy of reporting, analysis, and planning. Furthermore, it provides enough insight to assist the management to improve their operational performance, employee satisfaction as well as customer satisfaction. This will also have a cascading effect on their overall revenue to have increased and/or cost of production reduced when BI tools are used effectively within the organization.

6.4 Importance of HR Planning in Modern Industries

Manufacturing is considered as the backbone for the economic growth of a country. Under Make in India drive, the government has introduced several policy changes toward Ease of Doing Business, enabling new Startups, and supporting existing Micro, Small, and Medium Enterprises (MSMEs). The processes for registration of companies have been digitized, streamlined, and rationalized with respect to regulatory compliances to boost investor sentiments. Further, the revised Public Procurement Policy mandates that at least 25% of all annual public procurements should be done by MSMEs. The government has been investing in infrastructure development in terms of high-speed Internet, improved logistics, regulation, etc. Besides, organizations are adopting advanced technologies, and there is a significant shift in consumer behavior under digitally evolving market (Singh, Singh & Karki, 2021). Thus, it becomes more significant for organizations to make data-driven decisions and develop their HR to work under such progressing environment.

Human Resource Planning (HRP) is required for anticipation and planning of the current and future HR requirements of the organization as per their business goals. A HRP process ensures that the organization always has an adequate number of qualified persons at all times to perform the identified jobs of the enterprise (Arora et al., 2014).

The employment rate is on the tailspin from the day COVID pandemic has exploded across multiple countries of the world. Not only have job opportunities shrunk, but the nature of jobs has also changed dramatically in the post-COVID world. Greater digitization of workspaces, major shift in buying behaviors, and growth of digital services have engendered the need for multiple skills among employees.

The modern industries are under continuous challenge to shift their operational processes to meet the unprecedented disruptions caused by advancement in technology or pandemic conditions. The need for a workforce in terms of skills has become dynamic and unpredictable. Thus, modern industries are put to turbulent encounters of maintaining, predicting, controlling, and eliminating redundant workforce to ensure an optimum level of cash flows.

The process of HR planning becomes more important when more business variables are affected, such as customer buying behavior, customer requirements for product/service, availability of logistics, competitive pricing, limitation in operations due to pandemic restrictions, etc. BI is one of the innovative approaches that enables tailored HR planning for an organization's sustainability under dynamic and competitive conditions.

6.5 BI for HR Planning

HRP process is used by management to predict and induct new incumbents and reach the desired level of workforce to carry out the integrated plans of the organization (Pandey, 2016). It consists of projecting future workforce

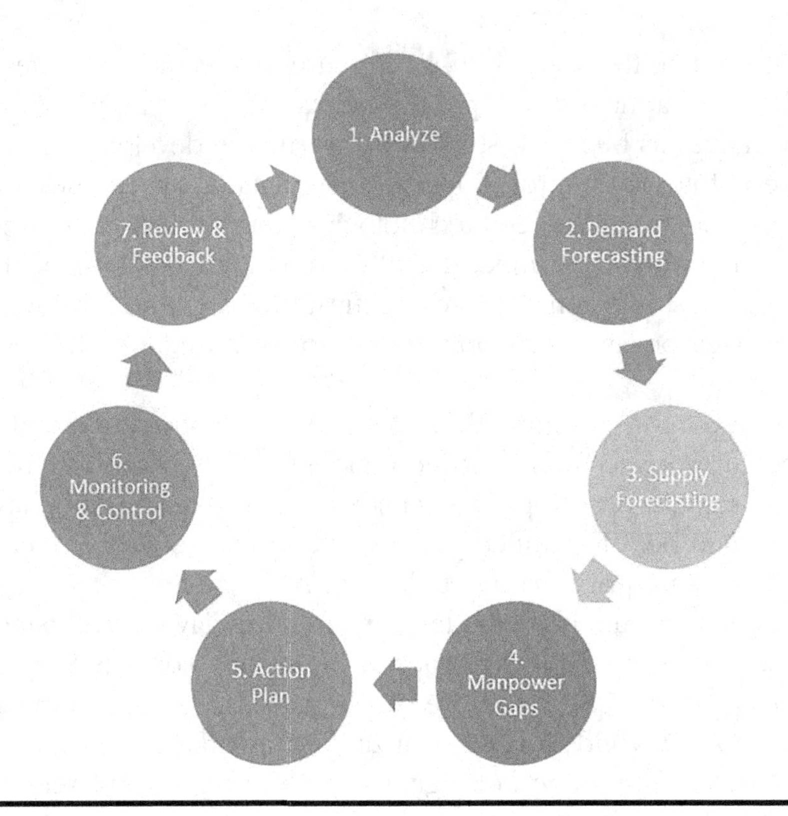

Figure 6.2 Process of HR planning.

requirements, developing workforce plans for the implementation of these projections. Most organizations align their activities as per the following steps of the HRP process (Figure 6.2).

6.5.1 Analyze Workforce Gaps (Steps 1–4)

Every organization analyzes its annual business plan and assesses the requirement for a workforce based on operations that existed, added, or changed to meet revised business targets (Step-1). The HR department of the organization defines the requirement in terms of job roles/positions, job description, critical factors, limitations, timelines, achievement targets, etc. (Step-2). Much of these activities require manual efforts and depend on intelligent capability of management or expert professionals. However, there are a few areas as mentioned below, where BI interventions enhance quality decision-making:

a. **BI for Supply Forecasting**: Presently, the outsourced recruitment agency helps the organization in predicting supply forecasting against their requirements (Step-3). However, to compete in the modern era, organizations must develop their own processes to make such predictions. This is feasible if the organization captures the potential applicants who are available for recruitment on a continual basis. This database of applicants is also known as Talent Pool. This database is a mix of people who have applied at a time when either the requirement was not there or applicant was not a good fit for that role with the company. Many organizations, such as Hyundai, Hindustan Power, USV Pvt Ltd., Max Life Insurance, etc., maintain landing pages on their websites where visitors are encouraged to drop their resume to consider for future requirements. Once such databases are available, the BI tools can be integrated with these databases of the organization. However, for effective predictions by BI system, it has been proposed that the applications may collect data/information on the parameters, such as Skill Name, Project and/or Client Name, Period, and Activity.

Once the database is available in terms of skills, it becomes easy for any BI Software application to match it with the requirement at any point of time of recruitment by the organization. Thus, a realistic supply forecasting can be made through such internal databases. These BI forecasts can also be used for scrutiny of applications. Such forecasts or predictions shall be based on other parameters, such as experience, type of project, activity or responsibilities undertaken, etc. Once the best suitable cohort of candidates is available, the organization can apply one of its conventional methods to select a candidate for advertised jobs.

A similar process can also be used to create an internal Talent Pool from existing employees of the organization who may have higher aspirational goals. This way, organizations will have more satisfied professionals at their disposal who are willing to contribute more by increased engagement and capabilities toward organizational goals.

b. **BI for Workforce Gaps**: At present, managerial judgment method is used to predict workforce requirements at a higher management level, whereas the work-study method is used to predict shop-floor requirements of HR. Further, the ratio-trend analysis, Mathematical Models, and other forecasting methods are also used. As per the prevailing practice, the consultancy organizations are mainly involved in workforce gap analysis. The starting point for such methods is the job, and it is defined by a regular job description that consists of job identification, job summary, job duties and responsibilities, working conditions, social environment, machines, tools and equipment, supervision, and relation to other jobs.

As most of the workforce gap analysis methods are scientific in nature, so, an organization can develop an IT-enabled system to capture such information. By adding a layer of tasks to each job and integrating with the BI application, the organization may be able to undertake workforce gap and skill gap analysis in real time. The broad component of additional layer may comprise task, frequency, and skill required or possessed by the candidate.

The BI system will be able to measure the performance of each individual based on task-related data in terms of workload assessments and workforce gaps, if any. The same structure can also be used for benchmarking job performance as well as building a framework for performance appraisal of an individual on the job.

6.5.2 Action Plan (Step-5)

As per the HR process, the HR department of the organization is entrusted with the responsibilities to develop action plan for training, procurement, transfer, retention, re-deployment, retrenchment, and re-employment (Step-5). The prevailing practices are mostly top-driven, where managerial judgment is mostly applied in making most of the decisions for the organization. Perhaps, such practices are known to be prone in developing errors of personal bias. Thus, appropriate business indicators can be identified and integrated with BI applications to overcome any kind of error and for better business decision-making as stated below.

a BI for External Procurement

Basically, there are two sources of recruitment: internal and external. The internal source consists of transfer, promotion, and recommendation by the existing employees, whereas external sources comprise advertisement, employment exchange, private employment consultants, campus interviews, rival firms, unsolicited applicants, etc. These traditional methods have transformed into innovative methods (Neelie, 2018), such as Job adverts (text-based advertisement), programmatic advertising (auto buying of digital space), video interviewing, benefitting from gig economy (freelancers for short-term), engaging passive candidates, employee referrals, texting, social media, virtual reality, employer review sites, aging workforce management, and mobile recruitment. The fundamental to all methods of recruitment is assessment of candidate's potential. Thus, the combination of Artificial Intelligence (AI) and BI should be used by industries to fulfill their requirement for recruitment. The BI will help in narrowing down the prospective candidates, whereas AI will fulfill the requirement of using advanced mathematical and statistical techniques to identify most potential candidates. Thus, organizations can become aware of the most suitable method for recruitment based on criticality, cost, and any other parameter relevant to the selection by BI system.

b BI for Internal Procurement

If the organization maintains a database of internal employees in terms of skills as well as holds details of tasks and skill sets required for all positions, it would be possible to apply BI to identify the best suitable candidate by transfer, promotion, re-employment, etc. Also, the database of successful candidates who were employed as interns, past performance records of employees, internal talent pool, etc., can be put together for BI to propose various trends and the probability of best fit. Further, the organization may also be able to frame retention policies based on fair negotiations that can be built after BI analysis and recommendations.

c BI for Training

After recruitment, the candidates are provided several training opportunities to build their capacity to meet the organizational requirements. There are several training provisions, such as orientation training, vestibule training, on-the-job training, management development, etc. Most of the organizations maintain limited information of such training with the personal records of employees, such as the title of the training program, duration, etc. But this is not enough to assess correctly the present or future requirements of the training needs for an existing employee. The requirements of jobs should be broken down into tasks and mapped with the skills possessed by the employees. The performance of the employees against these tasks and skills must be measured with the standard performance metrics. Such data, if available, will enable the organization to integrate with the BI tool to forecast skill gaps and help the HR department in identifying or designing capacity-building modules customized to the requirements of each individual to enhance the organizational capacity in a significant manner (Figure 6.3).

Figure 6.3 BI process for training.

6.5.3 *Monitoring and Control (Step-6)*

This step is mainly related to rules and regulations. The function also includes motivating the employees to optimally utilize their services for the maximum benefit of an organization. The controlling authority for each employee is defined by the hierarchical system of the organization. Presently, these authorities utilize various HR functions to monitor and control the organizational actions of these employees. A few benefits, such as compensation, leave, appraisal, promotion, retention, transfer, etc., are used to motivate employees. However, these benefits are subjective and can lead to disharmony if managerial judgment goes against moral and ethical acceptability of the recipient. Thus, another layer of soft skills needs to be captured by the HR department, where incidents are also recorded through AI tools. Subsequently, with the integration of BI applications, a comprehensive prediction can be made on motivating and demotivating factors for an employee, situations where employee will contribute to higher or lower output, etc.

6.5.4 *Review and Feedback (Step-7)*

This is the last and most important step for any organization. However, most of the organizations rarely give proper emphasis on recording and analysis of such information from employees. First, the organization needs to identify the objective and frequency of the feedback. The objective could be to improve organizational work culture, sensitization on gender issues, etc. Thus, modern organizations develop online systems where customized feedbacks are solicited from the existing employees at regular intervals and appropriate actions are undertaken timely so that employees remain motivated and highly contributive. The BI tools can analyze such data into comprehensive predictions to achieve the predefined goals for enhancing corporate performance.

6.6 BI Architecture and Components

A typical BI process is depicted in Figure 6.4, which can support business decision-making.

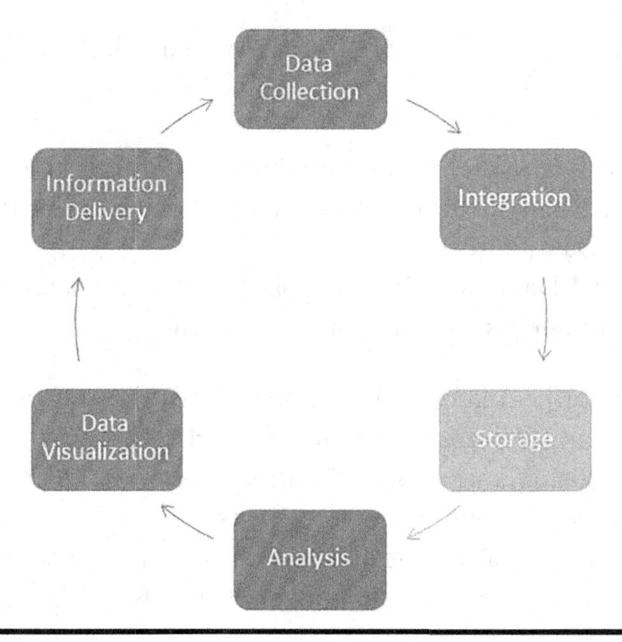

Figure 6.4 BI process.

The whole BI process is collectively supported by a set of independent core components. These components together make up a BI system that is deployed through cloud technology. These components are Data Management, Advanced Analytics, Business Performance, and Information Delivery (Kapoor, 2010). The basic architecture of the BI System is shown in Figure 6.5.

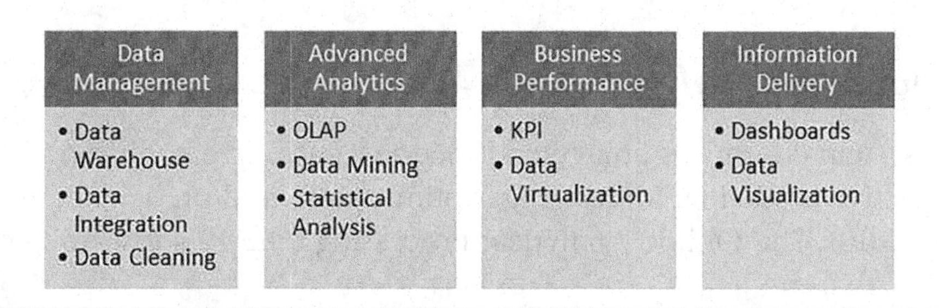

Figure 6.5 BI architecture.

6.6.1 Data Management

This subsystem manages the collection, storage, and structuring of data into databases in the following formats:

a Data Lake
Data Lake is a huge set of raw data that are stored in the native format for a purpose not yet defined.

b Data Warehouse
Data Warehouse is a storehouse for organized, filtered, and processed data.

c Data Mart
Data Mart is a subset of a data warehouse, but it holds data only for a specific department or line of business, such as sales, finance, or HRs.

d Operational Data Store
Operational Data Store is a snapshot of data gathered from multiple transactional systems for operational reporting.

The data are primarily collected from one or both the sources, i.e., (1) internal sources, such as software related to Enterprise Resource Planning, CRM, Finance, Manufacturing and Supply Chain Management, and (2) external sources, such as applications related to clients,
vendors, customers, etc. The criteria for selection of such sources are based on relevancy, frequency, data quality, level of details available, and data type.

The technology is built for data integration into a unified view. Such tools are known as Extract, Transform, and Load, where data are pulled into the predefined format and loaded into the BI System; OR Extract, Load, and Transform, where data are pulled and loaded as is and transformed later for specific BI uses. Besides, there are other Data Profiling and Cleaning technologies that are used to improve the quality of data after integration with data warehouses (Pratt, 2020).

6.6.2 Advanced Analytics

This subsystem comprises analytical functions, such as data mining, statistics, predictive modeling, forecasting, optimization, and predictive analytics (Kapoor, 2010). The Online Analytical Processing (OLAP) software allows the extraction of relational view from multiple databases, whereas software, such as SPSS, XLSTAT, multivariate data analysis, etc., are integrated with BI Systems for statistical analysis. Besides, other tools like data mining techniques are used to integrate data with external sources, such as data lakes, big data, or Hadoop, where data are unstructured. Data-mining techniques use algorithms to discover information automatically (Shmueli et al., 2010).

6.6.3 Business Performance

Since all BI Systems are data-driven, it becomes critical to identify the target indicators/business drivers for performance measurement of selected business processes. These indicators are also known as KPIs. Each KPI is a quantifiable parameter on which data are collected at regular intervals to assess the performance of the process or organization. The data analytical tools make other statistical assessments, such as correlation, variance, deviation, etc., to make predictions and recommendations to improve business decision-making. The technique of **Data Virtualization** is used by this subsystem to combine multiple data from different sources for comprehensive analysis based on pre-decided KPIs.

6.6.4 Information Delivery

The function of this subsystem is to provide an interactive user interface. This utilizes **Data Visualization** capability to eliminate unwanted data and ascertain patterns, trends, and insights. The tool on information delivery has built-in data visualizations capabilities and supports self-service for additional data analysis. As per Pratt (2020), BI dashboards and online portals provide real-time customizable views, whereas reports are more static. The visualization and interactive dashboards contribute to the market growth of BI applications.

6.6.5 Business Architecture of BI Application

The business architecture of BI application comprises components that facilitate the organization and structural aspect of data analysis and repository to store metadata (Pratt, 2020).

6.7 BI and Analytical Tools and Applications

The global BI adoption rate is 26% across organizations (Cartledge, 2021). On detailed analysis, it was found that organizations with a company size of 2,500–5,000 employees prefer BI solutions and put in place a central BI Team. Further, the BI landscape is constantly evolving and is getting a boost with the accessibility to technologies, such as Cloud computing, AI, etc. This adoption when planned in long term, the initial cost of the platform and its maintenance gets offset with the business benefits generated.

As many organizations have started adopting Big Data technology, the demand for BI and analytical tools has also improved significantly. This trend shall increase steadily with advancement and adoption of technology across the industries. As per BIM (2020) report, the global BI market will grow to USD 33.3 billion by 2025 at an annual growth rate of 7.6%. BI applications are likely to grow mainly for business functions, such as HR, Finance, Operations, Sales, and Marketing.

6.8 BI Adoption Strategy

Ministry of Health and Family Welfare (MoHFW), Ministry of Statistics and Program Implementation (MoSPI), Departments of Navy, Education, GSTIN, RTGS (Real Time Govt. Society), Niti Ayog, and State Governments of Rajasthan, Andhra Pradesh, and Assam are some of the government organizations that have onboarded BI Applications. Besides, private companies, such as LinkedIn, Amazon, Ferrari, Adobe, Cisco, Deloitte, and Walmart, are some of the larger enterprises that have successfully adopted it. Considering the large market size of MSME, it becomes a higher responsibility of the Service Providers as well as the Government to provide conducive conditions for market growth of the product (BI applications). Accordingly, the following three approaches are proposed:

6.8.1 Vendor Approach

As we all know, research and development attract substantial investment of a company on futuristic technology. The companies recover their costs of development of product or service by charging a higher fee as they hold the patents and copyrights to these innovations. These practices are prohibitory for deep penetration. However, with the increase in competition, such practices change to different market strategies from Rapid Skimming, Slow Skimming, to Rapid Penetration.

a. Under **Rapid Skimming**, the marketers launch the product at a higher price with high promotion. The price of the product is kept high to recover the cost of research and development. Efforts are being made to recover the research and development cost before competitors are able to bring a matching and similar product.
b. Whereas in **Slow Skimming**, marketers launch the product at a higher price with low promotions. Here, the promotions are avoided to recover maximum profit and minimize expenditure.
c. But in **Rapid Penetration**, marketers launch the product at a lower price with high promotions. This strategy is inspired from the commercial principle of "Economies of Scale," where price is kept low to capture maximum market share. In such cases, the cost of production is recovered by reaching out to the largest market size as well as pushing associated products or services, such as annual maintenance, etc.

6.8.2 Government Approach

The responsibility of the Government is more toward equal distribution of resources as well as providing conducive economic environment for protecting the smaller commercial entity from monopolistic practices of larger enterprises. Thus, in the present case of expansion of BI tools to modern industries, it becomes important for the Government to execute its role in the form of **Competition Watchdog**, where it looks on monopolistic malpractices motivated by pushing the products/services and maximizing profits to the disadvantage of the customer. To check upon them, the Government plays the role of **Regulator** to develop legal framework to provide protection to all entities involved in supply, procurement, and usage of such products or services. Besides, under special circumstances, when competitors are less and prices are higher, the Government also plays the role of a **Facilitator**, where it may subsidize the cost by absorbing part of it or it may develop capacity building of industrial strata that is unable to access such technology due to high cost. Further, in extreme cases, the Government can take the responsibility of providing basic functional system for a larger economic activity at an insignificant cost that can help the indigenous evolution of such systems.

6.8.3 MSME Approach

As it is apparent that these BI tools have mainly penetrated to the larger enterprises and with the government sector. Under COVID pandemic, the requirement of such tools has increased manifold with micro, small, and medium enterprises as well. These organizations are not much aware to such technologies. Besides, the cost of hiring expert BI professionals is too high for these organizations. Thus, the following basic MSME approach (Figure 6.6) is proposed to these organizations for making an informed choice in the selection of BI tools for themselves.

Figure 6.6 MSME approach.

a Market Analysis

Before adopting new technology, every organization should review its market through the Five Forces Model of Porter (Figure 6.7).

Figure 6.7 Five forces model of porter.

The Five Forces Model helps in predicting the level of competition within a certain industry. According to this framework, competition in an industry depends on five basic forces related to threat and bargaining power. Thus, based on the business strategy formed through Porter analysis, the organization may decide upon the technology to adopt that may include BI tools to their competitive advantage.

b BI Function

There are several BI functions (Figure 6.8) that are mainly required by an organization that may want to adopt BI tools to achieve business goals. It could be data mining, analysis of data, OLAP, MIS reporting, etc.

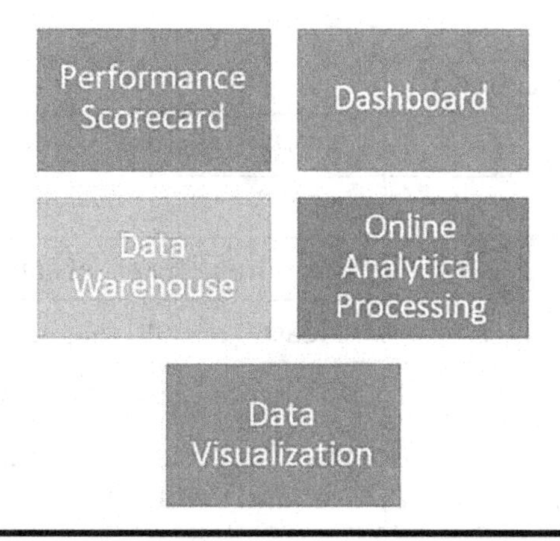

Figure 6.8 BI functions (indicative).

c BI Trends

No organization can run without BI. Previously, such expertise was available within the organizations and was dependent mostly on the capability of the executives. This approach was known as Traditional Method. However, with the transformation of the organization into digital hubs, more data are generated that is difficult for an individual to comprehend without technical tools, thus organizations have realized the potential of three BI trends, such as Mobile BI, Cloud BI, or Social BI. The mobile BI is mainly hardware-driven, where dashboard is available on the mobile phone, whereas in Cloud BI, the application is accessed through virtual cloud and it enhances scalability of BI solution. Whereas, Social BI is a kind of dashboard that is integrated with social media handles to publish business analytics reports.

d Deployment Strategies

Presently, there are two strategies to install these applications, i.e., on-premises at the data centers of the organization or to the virtual cloud from where the BI application is accessed by the organization.

The right deployment strategy can be selected based on some factors, such as the five Vs of Big Data and security.

e Competitive Landscape

There are many vendors that provide services related to BI tools. The prominent BI tools include SAP Business Objects, Zoho Analytics, Oracle BI, Domo, Datapine, MicroStrategy, SAS BI, Yellowfin BI, QlikSense, SiSense, Microsoft Power BI, Looker, Clear Analytics, Tableau, and IBM Cognos Analytics (Haije, 2019). Further, the following BI Selection Criteria is proposed in Table 6.1:

Table 6.1 BI Selection Criteria

Parameter	Criteria
Notable **features**	List
Pricing of the **product**	Not more than 2% of sales volume
Utility to the **organization**	At least it should map with five notable features
Sector specific rollout	List client organizations who have adopted this product
After sales service	Cost and feedback
Dashboards and user **experience**	Demo run / 90 days free trial
Other **issues**	If any

6.9 Conclusion

An established and modern HR management provides business insights to provide an optimum workforce with the help of modern HR tools. These insights are built on employee demographics, such as qualification, experience, training, performance, etc. As recruitment, training, and retaining majority impact the organization performance, thus selection of correct BI solution and objective is critical to modern industries. The BI is an advanced tool to enhance qualitative decision-making. The MSME approach drives effective implementation of BI tools at the organization level. However, this study is based on a consultative view based on informal discussions with experts. Further, there will be unknown challenges in the adoption and implementation of BI tools. Thus, a detailed comprehensive research should be undertaken with the existing clients who are using the services of BI tools.

References

Arora, N., Lasune, S., and Karande, V. (2014, November). *Business Management Paper-I (Book)*. University of Mumbai, https://old.mu.ac.in/wp-content/uploads/2014/04/Human-Resource-Management-Paper-I-English-Book.pdf.

Bagale, G.S., Vandadi, V.R., Singh, D. et al. (2021). Small and medium-sized enterprises' contribution in digital technology. *Annals of Operations Research*. Doi: 10.1007/s10479-021-04235-5.

Business Application Research Center (BARC). (2021). The benefits of business intelligence…why do organizations need BI? https://bi-survey.com/benefits-business-intelligence, (Accessed on April 13, 2021).

Business Intelligence Market (BIM). (2020, September). Business Intelligence Market by Component (Solutions and Services), Solution (Dashboards and Scorecards, Data Integration and ETL), Business Function (Finance, Operation), Industry Vertical (BFSI, Telecom and IT), and Region - Global Forecast to 2025. https://www.marketsandmarkets.com/Market-Reports/social-business-intelligence-bi-market-1048.html?gclid=CjwKCAjwm7mEBhBsEiwA_of-TMvBc0DMVk-OggY1Uxip5UgMmrXCDfuhBGlqsiLoEZqPFMljPRk9DxoCErcQAvD_BwE.

Cartledge, A. (2020). *Business Intelligence Trends, 360 Suite*. https://360suite.io/ebook/business-intelligence-trends-2020/ (Accessed on August 17, 2021).

Haije, E.G (2019). *Top 15 Business Intelligence Tools in 2021: An Overview*. Mopinion, https://mopinion.com/business-intelligence-bi-tools-overview/.

Kalchauer, A., Lang, S., Peischl, B., and Torrents, V.R. (2014). Business intelligence in software quality monitoring: Experiences and lessons learnt from an industrial case study. *Paper Presented at the International Conference on Software Quality*.

Kapoor, B.C. (2010). Business intelligence and its use for human resource management. *The Journal of Human Resource and Adult Learning*, 6(2), 21.

Kautish, S. (2008). Online banking: A paradigm shift. *E-Business, ICFAI Publication, Hyderabad*, 9(10), 54–59.

Kautish, S., and Thapliyal, M.P. (2013). Design of new architecture for model management systems using knowledge sharing concept. *International Journal of Computer Applications*, 62(11).

Mutongi, D.C. (2016). *Revisiting Data, Information, Knowledge and Wisdom (DIKW) Model and Introducing the Green Leaf Model*. Revisiting data, information, knowledge and wisdom (DIKW) model and introducing the green leaf model. *IOSR Journal of Business and Management*, 18(7), 66-71

Narayan, L. (2019). Employee engagement: How much has changed in India, in the past five years? HR Katha, https://www.hrkatha.com/special/employee-benefits-and-engagement/employee-engagement-how-much-has-changed-in-india-in-the-past-five-years/#:~:text=Engagement%20at%20work%20is%20highest, employees%20who%20were%20fully%20engaged (August 02, 2019).

Neelie, (2018). 13 innovative recruitment methods top recruiters use. harver.com, https://harver.com/blog/innovative-recruitment-methods/#GigEconomy (posted on July 19, 2018).

Pandey, S.C. (2016). An overview of human resource planning. *International Journal of Interdisciplinary and Multidisciplinary Studies (IJIMS)*, 4(1), 8–12. http://www.ijims.com/uploads/57daf1d942b7104c3eac3subs.pdf.

Pratt, M.K. (2020). Business intelligence architecture. TechTarget, https://searchbusinessanalytics.techtarget.com/definition/business-intelligence-architecture (accessed on 02 May, 2021).

Rowley, J. (2007). The wisdom hierarchy: Representations of the DIKW hierarchy. *Journal of Information Science*, 33(2), 163–180.

Shmueli, G., Patel, N.R., and Bruce, P.C. (2010). *Data Mining for Business Intelligence*. Wiley Publication.

Singh, A., and Gite, P. (2015). Corporate governance disclosure practices: A comparative study of selected public and private life insurance companies in India. *Apeejay - Journal of Management Sciences and Technology* 2(2).

Singh, D., Singh, A., and Karki, S. (2021). Knowledge management and Web 3.0: introduction to future and challenges. In *Knowledge Management and Web 3.0*. De Gruyter, Cambridge University Press. Doi: 10.1515/9783110722789-001Agents.

Tutunea, M.F., and Rus, R.V. (2012). Business intelligence solutions for SME's. *Procedia Economics and Finance*, 3, 865–870.

Chapter 7

The Current State of Business Intelligence Research: A Bibliographic Analysis

Rajeev Srivastava
University of Petroleum and Energy Studies

Sunil Saxena
Microsoft

Contents

DOI: 10.4324/9781003184928-7

7.1 Introduction

An increase in the number of academic publications due to emphasis on the empirical contribution produces articles in huge volumes and the diverse areas of particular themes or fields (Briner & Denyer, 2012). This makes the process of accumulating knowledge through the publication in previous research cumbersome. Scholars are using different types of quantitative and qualitative review processes for the literature review to understand and compile previous findings in the said areas. Bibliometric analysis is one of the best approaches for the systematic literature review process, which is based on statistical measurement and gives a reliable analysis process (Broadus, 1987; Diodato, 1994; Pritchard, 1969).

In this chapter, a bibliometric analysis using the R package bibliometrix is used to understand the literature review in the field of business intelligence (BI) to synthesize the findings of past research and use that existing knowledge base for further study in this field (Massimo & Corrado, 2017). The open-source software named the R language bibliometrix package is used for the analysis and graphical representation (R core team, 2016). This language also provides various graphical and statistical techniques along with the capacity for extension in the future (Matloff, 2011).

This chapter first presents the knowledge base of the research topic related to BI (Kautish, 2008; Kautish & Thapliyal, 2013), using descriptive analysis of the extracted documents. Second, the intellectual structure of the keywords is presented, and finally, the collaborative network structure of various authors, institutes, and countries is shown with the help of various tables and diagrams (Bagale et al., 2021).

7.2 Methodology

The bibliometric R-package is a collection of tools or functions used for bibliometric analysis. This package comprises various functions that are developed using R language. To perform the bibliometric analysis, 1,540 documents were finally selected from the Web of Science (WoS) database between the duration of 2016–2021 (July). This step allows the scholar to create his/her database (Waltman, 2016). While applying filter the journals in various fields, including Management, Business, Social Sciences Interdisciplinary, and Economics, were included assuming these are the journals related to management and related fields (Singh, Singh & Karki, 2021). During filter, the type of document "Articles" is included, which are published in English language. After selecting 1,540 articles, these were exported in the form of text files. These text files are imported in biblioshiny graphical user interface developed in R-language. Finally, a bibliometric analysis is performed on the text files to get the main information related to the document extracted (Kautish, Singh, Polkowski, Mayura & Jeyanthi, 2021). The descriptive network and collaborative analysis are performed to extract the contributions of various authors, countries, and journals in the field of BI.

7.3 Analysis

Various types of functions are used for the descriptive analysis of the data related to various articles on some specific theme or topic. The main measures of these data are analyzed using the bibliometrix analysis function, which is a list of various elements, such as articles, authors, affiliations, citations, and many more. The functions, such as summary and plot, are used to summarize the main results obtained using bibliometric analysis. The main results of the bibliometric analysis from the bibliometric data frame are presented in different tables and figures.

7.3.1 Descriptive Analysis

The purpose of descriptive analysis is to get the information related to the number and types of a document collected for analytics. It also involves the analysis of relevant sources and authors who have contributed to the field of BI.

The main information related to the collection of data used in the bibliometric analysis is presented in Table 7.1. This indicates a total of 1,540 articles considered in this bibliometric analysis from the duration of 2016–2021 (July). A total of 4,822 authors contributed in this field with average citations of 8.6 per article.

The details of the number of articles published in the previous 5 years from 2016 to 2020 is summarized in Table 7.2, which is a clear indication that the number of publications is increasing year by year except for 2019. The trend shows that in the coming years the number of publications will increase.

Table 7.1 Main Information

Description	*Results*
Timespan	2016–2021 (July)
No. of sources	778
No. of articles	1,540
Average years from publication	2.79
Average citations per document	8.6
Average citations per year per doc	2.15
Authors contributed	4,822

Table 7.2 Productivity of Articles

Year	*Articles*
2016	250
2017	283
2018	315
2019	276
2020	347
2021 (July)	41

Many authors have contributed in this field. The details of the relevant authors are summarized in Table 7.3. According to analysis Fihn, S. D., is the most impactful author with an h-index of 9. There are 11 authors with h-index either equal to or above 5.

During the analysis, it has also been tried to know the countries having a maximum contribution of published articles in this field, the details of which are summarized in Table 7.4. The United States (n=1,076) produces the maximum number of articles. China produces (n=696) articles in this field. A total of 10 countries have contributed more than 100 articles in this field.

Analysis of global citations for different documents was done to understand which documents have the maximum number of citations. The results are summarized in Table 7.5 and indicate that the document "Levine G. N., 2016, CIRCULATION" has a maximum citation followed by "GILBERT T, 2018, LANCET."

Table 7.3 Most Relevant Authors

Author	*H_Index*
Fihn SD	9
Chen XH	8
Popovic A	8
Hu CH	6
Wang HS	6
Li Y	6
Ghasemaghaei M	6
Abello A	5
Daneshmand M	5
Chang JY	5
Romero O	5

Table 7.4 Top Ten Most Relevant Countries

Region/Country	*No. of Articles Published*
USA	1,076
China	696
Australia	314
UK	195
Spain	161
Germany	142
Canada	141
India	131
Italy	112
The Netherlands	108

The affiliation statistics related to the top universities producing the maximum number of articles are presented in Table 7.6. The most dominated university found during the analysis is "Hunan Univ Commerce."

Information related to the most relevant sources of the articles in the form of the journals is summarized in Table 7.7. The journal name *Journal*

Table 7.5 Most Globally Cited Documents

Paper	*Total Citations*
Levine GN, 2016, *Circulation*	660
Gilbert T, 2018, *Lancet*	215
Cai HM, 2017, *IEEE Internet Things*	181
Verma S, 2017, *IEEE Commun Surv Tut*	158
Al-Ali AR, 2017, *IEEE T Consum Electr*	152
LEAHY MF, 2017, *TRANSFUSION-A*	149
Qi LY, 2018, *Future Gener Comp Sy*	132
Horne BD, 2018, *Am J Resp Crit Care*	109
Rehman MHU, 2016, *Int J Inform Manage*	104
Greenwood BN, 2017, *MIS Quart*	104

Table 7.6 Most Relevant Affiliations

Affiliations	*Articles*
Hunan Univ Commerce	77
Univ Washington	37
Univ Melbourne	36
Cent S Univ	35
Southwestern Univ Finance And Econ	28
Univ Zagreb	28
Univ Arizona	24
Wroclaw Univ Sci And Technol	22
Stevens Inst Technol	21
Univ Ljubljana	21

of Intelligence Studies in Business is the top journal having the maximum number of articles published.

7.3.2 Network Analysis of Publications

The extracted document has various attributes, such as author, journal, citation, country, affiliation, and many more. These attributes are connected,

Table 7.7 Most Relevant Sources

Source Name	*No of Articles*
Journal of Intelligence Studies in Business	29
IEEE Access	23
International Journal of Information Management	23
International Journal of Advanced Computer Science and Applications	22
Sustainability	22
Decision Support Systems	17
Industrial Management & Data Systems	14
Journal of Computer Information Systems	13
IEEE Internet of Things Journal	11
Journal of Decision Systems	11
Sensors	11
Expert Systems With Applications	10
International Journal of Production Research	10

for example, author is connected to journal, keyword is connected to date of publication. In bibliometric analysis, these connections are shown with the help of a matrix form between document and attribute (document×attribute). All these networks can be easily identified and visualized using bibliometric analysis (Kamada & Kawai, 1989).

Co-word analysis is used to extract the most important keywords from the document to understand the conceptual structure of the research in that field (Callon, Courtial, Turner & Bauin, 1983). Based on this study, the most frequently used keywords are identified from the titles of published articles. The details of the most frequently used keywords are presented in Table 7.8. The result shows that the word "BI" is the most frequent because it is the keyword used to extract articles. The result shows that the word "Management " and "Model" are the two most frequently used words in articles. This indicates that BI is used by different organizations to develop better models and performance.

The details of the keywords used by authors most frequently in different articles is shown with the help of word cloud in Figure 7.1.

Table 7.8 Most Frequent Words

Words	*Occurrences*
Business intelligence	201
Management	155
Model	122
Performance	115
Big Data	101
Impact	100
Systems	93
Analytics	90
Information	82
Framework	65
Technology	61
Design	50
Information Technology	50
Knowledge	47
Quality	47
Success	45
Intelligence	44
Information systems	38
Challenges	36
Care	35

7.4 Collaboration Network of Authors, Institutes, and Countries

Bibliometric analysis is performed to determine the collaborative research work done between various authors, countries, and institutes that have contributed to the field of BI. In this section, the collaboration network diagrams along with the details of collaboration are presented.

7.4.1 Collaboration Network of Authors

The collaboration network between various authors is shown in Figure 7.2. During analysis, four clusters are identified, which work in collaboration

Figure 7.1 Word cloud of authors' keywords.

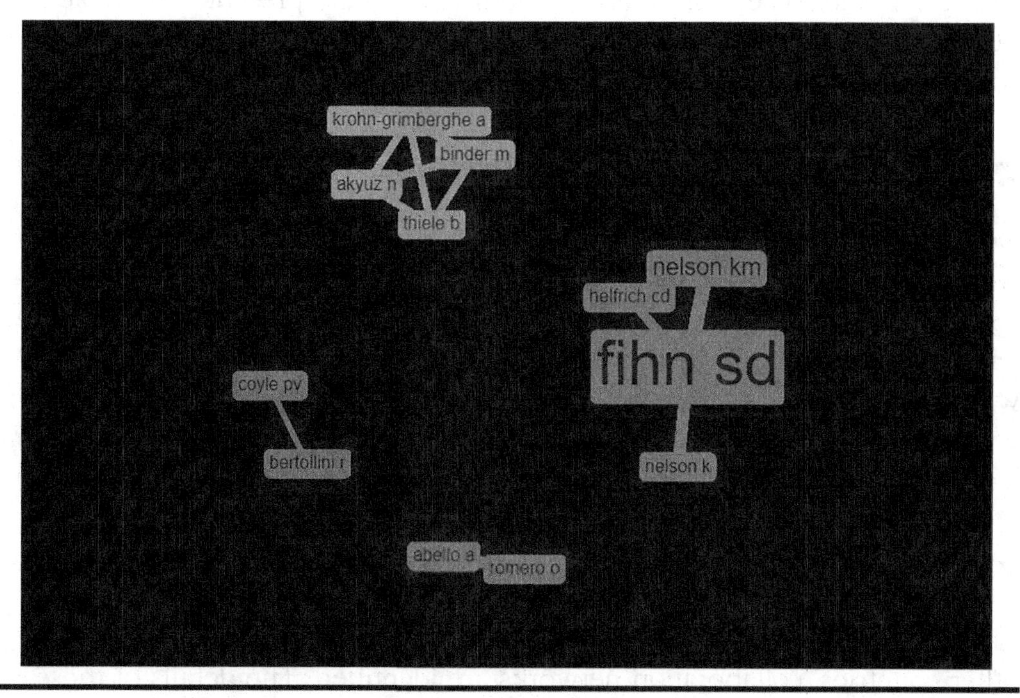

Figure 7.2 Collaboration network of authors.

Table 7.9 Cluster of Collaboration Network of Authors

Node	*Cluster*
Romero O	1
Abello A	
Nelson K	2
Fihn SD	
Helfrich CD	
Nelson KM	
Thiele B	3
Akyuz N	
Krohn-Grimberghe A	
Binder M	
Coyle PV	4
Bertollini R	

with each other. The details of these four clusters are presented in Table 7.9. Author name "Fihn SD" is found to be the author conducting the maximum collaborative study with "Nelson KM."

7.4.2 Collaboration Network of Institutes

During the analysis, it has been found that many Institutes/Universities are working in collaboration with each other. The details of the institutes working in collaboration are shown in Figure 7.3. The figure shows that countries having institutes, such as "Hunan Univ Commerce," work in collaboration with "Cent S Univ." A total of 10 clusters are identified, which are working in collaboration with each other. The details of these 10 clusters is presented in Table 7.10.

7.4.3 Collaboration Network of Countries

To know the most relevant countries conducting collaborative studies with other countries, collaboration networks are identified. The details of these

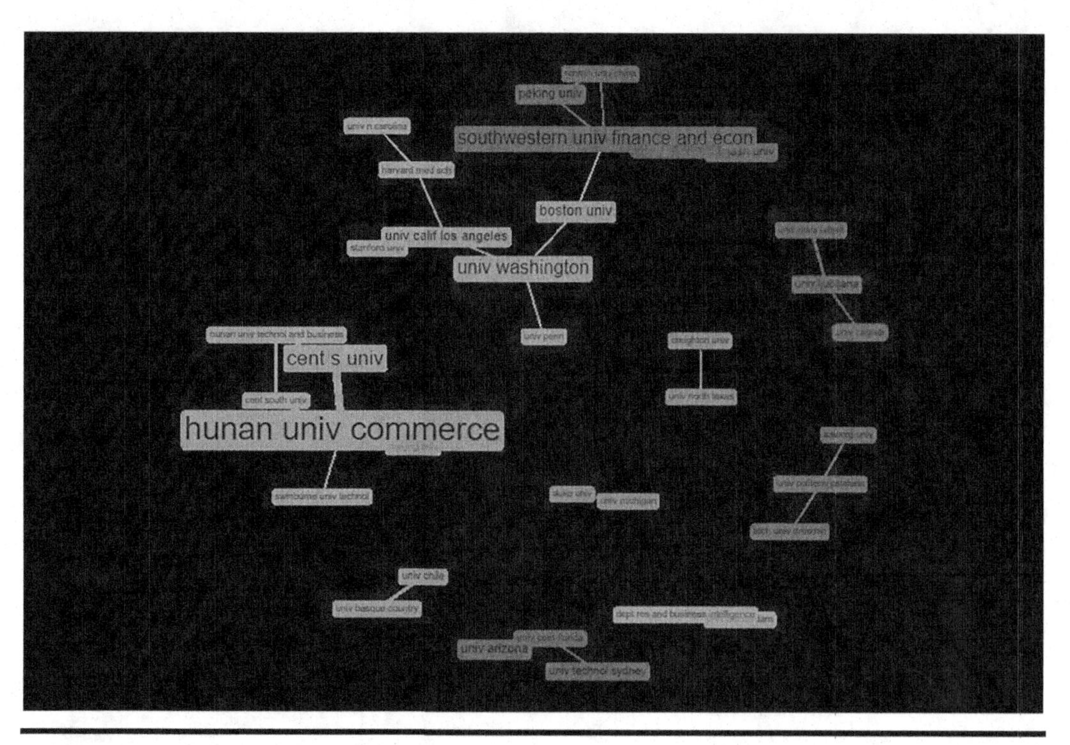

Figure 7.3 Collaboration network of institutes.

Table 7.10 Cluster of Collaboration Network of Authors

Node	*Cluster*
Univ Zagreb	1
Univ Ljubljana	
Univ Nova Lisboa	
Aalborg Univ	2
Univ Politecn Cataluna	
Tech Univ Dresden	
Duke Univ	3
Univ Michigan	
Univ Technol Sydney	4
Univ Cent Florida	
Univ Arizona	

(*Continued*)

Table 7.10 (*Continued*) Cluster of Collaboration Network of Authors

Node	*Cluster*
Univ Chile	5
Univ Basque Country	
Monash Univ	6
Univ Melbourne	
Renmin Univ China	
Southwestern Univ Finance And Econ	
Peking Univ	
Swinburne Univ Technol	7
Nanjing Univ	
Hunan Univ Commerce	
Cent S Univ	
Hunan Univ Technol And Business	
Cent South Univ	
Univ North Texas	8
Creighton Univ	
Univ Amsterdam	9
Dept Res And Business Intelligence	
Univ Washington	10
Univ Penn	
Boston Univ	
Stanford Univ	
Univ Calif Los Angeles	
Harvard Med Sch	
Univ N Carolina	

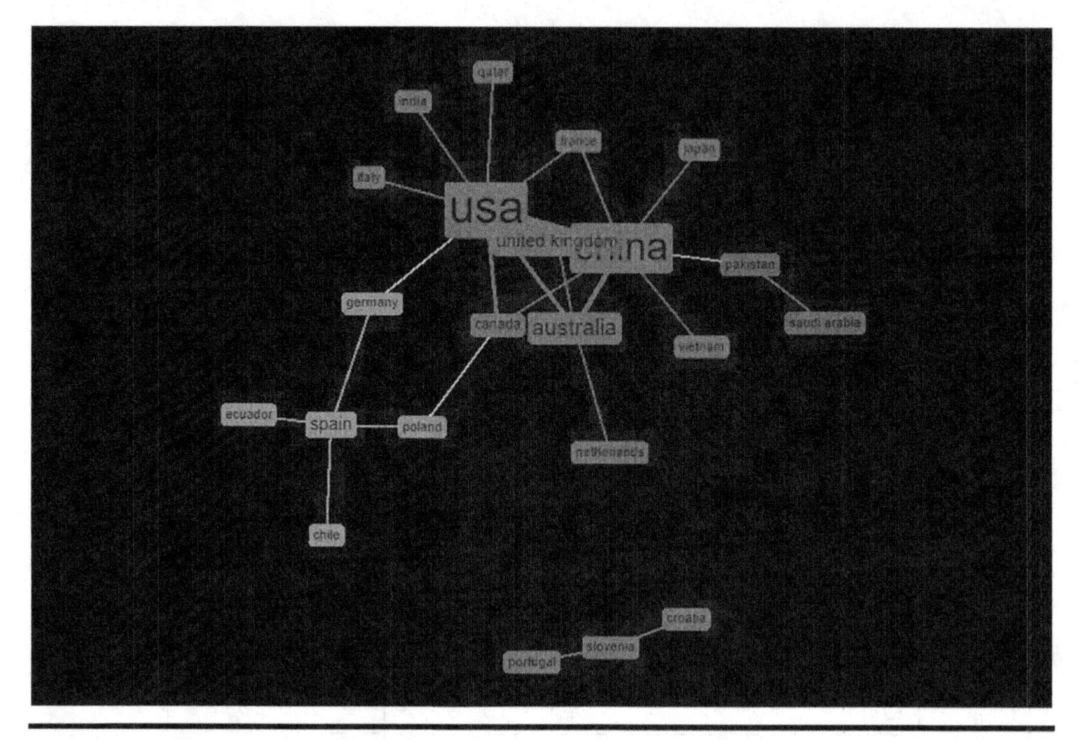

Figure 7.4 Collaboration network of countries.

networks are shown in Figure 7.4. A total of four clusters have been identified in the details, which are summarized in Table 7.11. In cluster 1, the United States and China are conducting maximum collaborative studies. Canada and Australia are also conducting collaborative studies with these two countries. Similarly, in cluster 3, Germany and Spain are conducting collaborative research.

7.5 Conclusion

A bibliometric analysis using the R-package Bibliometrix is used to understand the literature review in the field of BI to synthesize the findings of past research and use that existing knowledge base for further study in this field. To perform the bibliometric analysis, 1,540 articles were finally selected from the WoS database between the duration of 2016 and 2021 (July). Various

Table 7.11 Cluster of Collaboration Network of Countries

Node	*Cluster*
USA	1
France	
Australia	
China	
India	
Canada	
United Kingdom	
Japan	
Italy	
The Netherlands	
Vietnam	
Qatar	
Croatia	2
Slovenia	
Portugal	
Poland	3
Germany	
Spain	
Ecuador	
Chile	
Pakistan	4
Saudi Arabia	

types of functions are used for the descriptive analysis of the data related to various articles on some specific theme or topic. The study shows that the number of citations is increasing year by year. The journal named *Journal of Intelligence Studies in Business* is the top journal in this field. The "University of Nevada" is the most relevant and produces the maximum number of articles. The document "Levine G. N., 2016, CIRCULATION" has a maximum citation followed by "GILBERT T, 2018, LANCET." During the analysis, to know the countries having a maximum contribution to publish, it has been found that the United States and China produce the maximum number of articles. Network analysis shows that the words "management" and "model" are the two most frequent words used in articles. Collaborative analysis between various authors, institutes, and countries found that the author name "Fihn SD" conducting the maximum collaborative study with "Nelson KM". Countries with institutes, such as "Hunan Univ Commerce," are working in collaboration with "Cent S Univ." A total of 10 clusters are identified, which have been working in collaboration with each other. Countries, such as the United States and China, are doing maximum collaborative studies.

7.6 Future Direction and Limitation

This study identified the most relevant authors, journals, and countries that contributed to the field of BI. However, in which specific areas they have contributed will be the area of future research. Similarly, the various networks and collaborations between keywords, authors, universities, and countries are identified, but the details of their contributions are yet to be identified. From a limitation perspective, this study includes only the WoS database. In a further study, one can collaborate with various databases, such as Scopus, to get more detailed knowledge in this area.

References

Bagale, G.S., Vandadi, V.R., Singh, D. et al. (2021). Small and medium-sized enterprises' contribution in digital technology. *Annals of Operations Research*. Doi: 10.1007/s10479-021-04235-5.

Briner, R. B., & Denyer, D. (2012). Systematic review and evidence synthesis as a practice and scholarship tool. In D. Rousseau (Ed.), *The Oxford Handbook of Evidence-Based Management: Companies, Classrooms, and Research* (pp. 112–129). (Oxford Library of Psychology). Oxford University Press.

Broadus, R. (1987). Toward a definition of bibliometrics. *Scientometrics*, 12(5–6), 373–379.

Callon, M., Courtial, J.-P., Turner, W. A., & Bauin, S. (1983). From translations to problematic networks: An introduction to co-word analysis. *Social Science Information*, 22(2), 191–235. Doi: 10.1177/053901883022002003.

Diodato, V. (1994). *Dictionary of Bibliometrics*. Binghamton, NY: Haworth Press. http://www.bibliometrix.org.

Kamada, T., & Kawai, S. (1989). An algorithm for drawing general undirected graphs. *Information Processing Letters*, 31(1), 7–15 [Elsevier].

Kautish, S. (2008). Online banking: A paradigm shift. *E-Business, ICFAI Publication, Hyderabad*, 9(10), 54–59.

Kautish, S., Singh, D., Polkowski, Z., Mayura, A., & Jeyanthi, M. (2021). *Knowledge Management and Web 3.0: Next Generation Business Models*. Berlin: De Gruyter.

Kautish, S., & Thapliyal, M.P. (2013). Design of new architecture for model management systems using knowledge sharing concept. *International Journal of Computer Applications*, 62(11).

Massimo, A., & Corrado, C. (2017). Bibliometrix: An R-tool for comprehensive science mapping analysis. *Journal of Informetrics*, 11(2017), 959–975.

Matloff, N. (2011). *The Art of R Programming: A Tour of Statistical Software Design*. San Francisco, CA: No Starch Press.

Pritchard, A. (1969). Statistical bibliography or bibliometrics. *Journal of Documentation*, 25, 348.

R Core Team. (2016). *R: A Language and Environment for Statistical Computing*. Vienna, Austria: R Foundation for Statistical Computing. https://www.R-project.org.

Singh, D., Singh, A., & Karki, S. (2021). Knowledge management and Web 3.0: Introduction to future and challenges. In *Knowledge Management and Web 3.0*. De Gruyter, Cambridge University Press. Doi: 10.1515/9783110722789-001Agents.

Waltman, L. (2016). A review of the literature on citation impact indicators. *Journal of Informetrics*, 10(2), 365–391.

Chapter 8

Empirical Assessment of Artificial Intelligence Enablers Strengthening Business Intelligence in the Indian Banking Industry: ISM and MICMAC Modelling Approach

Sameer Shekhar
Indian Institute of Foreign Trade (IIFT)

Chandan
CHRIST (Deemed-to-be University)

Contents

DOI: 10.4324/9781003184928-8

8.1 Introduction

Enterprises are heading towards growth and their sustenance in the modern world, which is completely driven and ruled by data. Technology and innovation are using these data and big data for the purpose of reaching inference on the dataset. The human intelligence is being used and anticipated by AI using machine algorithm leading to the trends identification and getting more trusted insights strengthening business intelligence (BI) in an organisation enabling in making faster decision.

In today's dynamic and fast-changing business environment, the traditional flow of information in any organisation will not fulfil the requirements of top-level management, middle level management, and even lower-level management due to possibility of inaccuracy of information or delay in its dissemination. Also, the obsolete system fails to deal with the larger number of data and information. The introduction of BI is believed to provide solutions to these issues (Turban et al., 2008) processing huge amounts of data helping in better decision-making. BI is recognised as one of the significant aspects in an organisation's decision-making by means of collecting, classifying, integrating, analysing, and presenting real-time data (Inmon, 2002) and business information along the technological applications and automation practices and thus is called Data-driven Decision Support System (DDS). In the banking system, the use of BI plays a vital role in providing inputs to the banking system about the customers' decisions, based on which the industry can take a decision with regard to product (scheme) designing and as such. The technological innovation has brought advancement in business practice by providing enough scope for the implementation of Artificial Intelligence (AI) and Machine Learning.

The paradigm of BI (Kautish, 2008, Kautish and Thapliyal, 2013) is shifting towards AI as the former one is facing challenges of unstructured data unspecified analysis metrics and further not the unsupported visual representation. Digital Transformation is one of the most critical drivers on how companies will continue to deliver value to their customers in a highly competitive and ever-changing business environment (Bagale et al., 2021). AI has been recognised as one of the central enablers of digital transformation in several industries. As far as the implication of AI in banking and its impact on BI is concerned, it is easily known fact that the former one has great influence over the latter one and thus plays a vital role providing inputs to the management in taking business decisions on the propositions and predictions made by AI.

As the chapter deals with the assessment of AI enabling BI in the Indian Banking system, it is essential to understand the concept of AI, which functions with the primary objective of modelling of human behaviour, thinking and intelligence, and thus based on that, making a rational decision. In business nowadays, the implementation of AI amidst BI has become a top priority to deal with the growing customer base and associated to banking industry queries forming big data. The adoption of AI has been increasing across the service sector and holds paramount importance in the development of sound marketing strategy resulting in the expansion of customer engagement through strengthening BI. Extensive usage of AI has been reported to impart not only better understanding of service offerings' characteristics, types, and benefits to the users but also has been found helpful in grievance redressal and managing their queries systematically and enriching the ground of BI, which provides ample scope for academic research in this field. There have been various factors limiting and driving the implementation of AI concerned with information quality and information systems affecting BI and thus customer engagement.

8.2 Literature Review

Banking has been a congruent sector for considering innovations and technological upgradation (Shu and Strassmann, 2005) as those which are adopted by the banking institutions have been found helping in enabling new communication channel. The technological innovations in banking are not just limited to enriching the communication and coordination with customers and among different office levels (Hwang et al., 2004) but are also concerned with the data analysis technique being used for risk evaluation (Huang et al., 2011) and fraud detection (Ngai et al., 2011) purposes. BI is perceived as one of the most significant technological implications in banking sector as it is applied for decision-making through data mining, data warehousing, and data processing pertaining to credit evaluation, performance of the bank, internet banking, segmentation, and customer retention (Moro et al., 2015). Earlier, the non-technical users had to depend on a centralised data system to get their queries solved and faced a delay issue between the input (queries) and output (insights). In today's highly competitive and dynamic business environment, the banking institutions need to have a well-designed intelligence system that may help to sustain the challenging situation by proving output to make wise management decisions.

It is a data-driven intelligence model for predictive analytics and provides managers with the most probable outcomes (Zhang et al., 2017; Najmi et al., 2010; Rao and Kumar, 2011; Sundjaja, 2013; Xia and Gong, 2014) along the development of a mathematical model by extracting, transforming, and analysing large datasets (Fitriana et al., 2011). As a huge customer base is there seeking financial products and banking services, a huge dataset is the subject of dealing for the banks, which if required to be carried out traditionally, it would have been quite difficult to manage the system and serve the customer with promptness as is being done today along technological applications. Reforms in the financial sector in economies across the globe have led to multiple fold responsibilities and an increase in its operational network. It has led to credit extension and credit risk, customer base, and complexity as well, less dependency on traditional methodology and new scope of fraudulence, etc.

Enhancement in the customer base necessitated the use of disruptive technology, i.e., AI implications in the banking sector to serve the customers and thereby help the management in making predictions based on human inputs to the system as it learns to behave similarly as humans do and converts the computer smart enough into AI. In the era of drastic change in a business scenario, it has emerged as disruptive technology in the digital ecosystem and is believed to shape up BI. AI implication in the banking sector is being introduced in India by almost all the banking institutions, which caters customer has been found helping in customer query focussed, customer management focussed, operation focussed, finance focussed, and fraudulence focussed issues at front office, middle office, and back office levels. The AI has been gathering momentum recognising that the traditional analytics as a BI component is not effective enough to comply with the present day need of quick intelligence amidst big data. The alliance of AI with BI is bringing a huge difference in decision-making out of big data on human intelligence. Considering the same, it becomes important to identify the factors acting as enablers of AI implications towards stronger BI for rational decision-making for the management about different financial affairs, organisational operational mechanisms, customers' willingness, and their management.

Several research revealed different sets of enablers to AI implication in different businesses have been listed and presented in Table 8.1.

Based on the literature reviewed and accessed by the researchers, it has been found that there is very less study conducted on the modelling of the enablers of AI implications further strengthening BI implications in the banking sector.

Table 8.1 Select Enablers of AI Implication in Different Businesses

Author	*Variables*
Naumann and Rolker (2000)	Reputation, objectivity, variability, accuracy, completeness, response-time, reliability, and latency
Delone and McLean (2004)	Relevance, completeness, accuracy, and dynamic content
Hussein et al. (2007)	Accuracy, comprehensiveness, completeness, and timeliness
Ayyash (2015)	Accuracy, relevancy, timeliness, and completeness
Zohuri and Rahmani (2019)	User-friendliness, compatibility, consistency, timeliness
Mariemuthu (2019)	Information technology infrastructure, cost, competitive pressure, regulation and mimetic pressure, new technology competence
Sandkuhl (2019)	Compatibility, user-friendliness, training, and information
Munir and Rahman (2015)	Accuracy, reliability, timeliness
Jovovic et al. (2016)	Stability and reliability
Seddon (1997)	Bugs in the system, consistency of user interface, ease of use, documentation quality, programming
Calisira and Calisirb (2004)	System capability, compatibility, perceived ease of use, flexibility, user guidance, learnability, minimal memory load, perceived usefulness
Wixom and Todd (2005)	Reliability, flexibility, integration, accessibility, timeliness
Castelli et al. (2016), Elgammal (2019)	System programming, advanced algorithm, timeliness
Hameed (2018)	Modern-looking interface, reliability, flexibility, user-friendliness
Larivière et al. (2017), Solberg (2020), Montargot and Lahouel (2018)	Comfort of employee with digital technology use, knowledge about digital platform, compatibility

(*Continued*)

Table 8.1 (*Continued*) Select Enablers of AI Implication in Different Businesses

Author	*Variables*
Goodhue and Thompson (1995), Alwan and Al-Zubi (2016), DeLone and McLean (1992), Hartwick and Barki (1994), Kumar (2017)	Good system quality, reliability, flexibility, system capability, compatibility, accessibility
Jun and Kang (2013), Lin and Chang (2011)	Good AI information system, accuracy
Lewis et al. (2003), Smith and Tushman (2005), Aladwani (2001)	Top-level commitment towards uses of technology

8.3 Identification of the Enablers for ISM Modelling

For the purpose of conducting the study, the enablers depicted in Table 8.1 were listed and put before the banking professionals and AI professionals who helped in the identification of the select enablers relevant to the banking sector and were confirmed about its relevance to strengthen a strong BI and wiser decision-making ability of the management.

New Technology Competence: There is still a sense of reluctance among the considerable population chunk to use the internet banking and plastic card-based money transaction, which shows how unsecured the financial customers are there in India. Many people avoid using the banking operation on the gadgets in their hand. The AI-based chat bots or robots are not being used on a larger scale by Indian customers to enquire or for the purpose of grievance redressal. The less technology competence is one of the major impediments in meeting the objective of an organisation behind the implementation of AI in the organisation.

Training and Skill Development: The financial customers in India come from different demographic backgrounds, namely with different economic capacity, income figures, educational background, understanding power, different ages, etc., who may not be well versed in making use of the AI applications. Not only the customers, but also the employees need to be skilled and trained enough to get acquainted with AI implications in the banking sector at front office, back office, and middle office management. Therefore, to enhance their knowledge on the use of technology, type of information, which can be fetched using modern day automation, how to use, etc., the training and information dissemination will play greater role.

Top-Management Alignment and Commitment: AI is adopted at front desk, middle desk, and back office in banking and financial system. The top-level management needs to put in practice the use of AI in all the three dimensions with a stronger sense of commitment. The alignment of AI in front, middle, and back desk will provide validated and more accurate information to the BI system helping in taking rational decision.

Wide Use of AI Technology (Chat Bots/Robots): For more effective decision-making through standard DDS it is very essential to gather data through AI systems on a larger scale. The uniformity of the big data will provide scope for DDS to analyse and present the outcome for more acceptable inference.

Financial Support to Adoption: Financial support plays vital role in the adoption of any advanced bit of technology by an organisation. In India, still there are banks that have not sifted substantially towards the use of AI for customer management. Non-competence of the customers is natural to happen when it comes to the use of AI technology for making their queries even, but the adoption of the technology in modern day is a must for any organisation. Keeping these facts in consideration, it can be emphasised that the organisation should extend sufficient financial support.

Comfort of Employees with Digital Technology: Not only the customers in the banking sector, but also the employees should be well versed and skilled to use the digital technology (AI) in the middle office and back office for administration, Anti-Money Laundering, performing regulatory check of Know Your Customers (KYC), settlement clearance, record maintenance, accounting, information technology handling, etc. The competence of employees to the use of AI will help to gather more productive data to form a stronger BI.

User-Friendliness of the System Interface: The properties and structure of any technological system define the extent and frequency of its use. A system should be designed in such a way that does not make the user feel complex. Many studies suggest that a system should be quite comprehensive and easy to use, so that the objective behind the implementation may be fulfilled.

Reliability of the Information Given by AI: It is concerned with the dependability on the information and its tendency to deliver against customers enquiry that seeds trust in them towards the organisation or system (Rexha, Kinsgshott and Shang, 2003). It makes the system promising to the customer (Yang and Fang, 2004), and therefore plays a vital role in influencing customer satisfaction. The reliability of AI information is an important criteria in determining the level of user satisfaction with the AI used in banking.

Advanced Algorithm: AI is based on algorithms, which are responsible for the analysis of data, and to make predictions based on human intelligence inputs. As the age of technology is operating on data, businesses are using AI that relies on stronger machine algorithms to decode human needs and behaviours based on which BI tends to make wise decisions for the company.

Consistent Data: The number of customers are increasing at an unprecedented pace due to an increase in the population, economic activity, and economic growth, and thus the financial institutions and banks have to deal with the flooded information. The consistency of data under such circumstances will not lead to any ambiguity and thus will lead to more intelligent output. Lack of consistency in information provided by the users creates problems in releasing the right information about certain segments. Consequently BI may not work at it best to help the organisation in making rational decisions.

Multiple Query Management Capacity: The AI platforms, such as chatbots and robots, are installed with all the possible queries for the party for which the system has been designed. The flexibility of information is a very important aspect in AI implementation, especially in front office customer management in the banking sector. It is done based on industry specific market/ customer survey as well as the employee' experience at physical front office. The information required by customers about the financial products and services have been increasing day by day, therefore, the AI platforms are required to be updated with the relevant queries, which can answer the customers with accuracy and thus, based on that BI may come up more strongly in decision-making.

Good AI System Quality: The quality of AI system depends upon several elements such as language, accuracy, responsiveness, accessibility, programming, algorithm, bug freeness, etc. If all these determinants are present in a system, there is a huge probability of fair data from customers. The ease and comfort and all the mentioned characteristics define the customer's choice and lean towards the use of the AI platform for the purpose with which it has been designed and launched. Errors in the system, inaccuracy, inconsistency, weak programming, etc., will not allow the organisation in getting the set target, which will not support BI system in processing the data and coming out with right decision.

Good AI Information Quality: Quality of information decides the quality of data management, processing, analysis, and thus decision-making. The goodness of AI information quality is defined by several factors such as latency, language, comprehensiveness, reliability, consistency, relevance, timeliness, etc. Possessing these properties, it sets BI in a stronger state to have more particular in decision-making along human intelligence and predictions.

Rational Big Dataset: Decision-making is the most significant mechanism of a business organisation as it leads to the sustenance, growth, or loss in a business. A strong BI helps in taking wise decision based on past and present data trend pertaining to the business operation. In an organisation, BI allows measurement, processing, and management of big data and in having deeper understanding so that the company may take proactive decisions towards customer satisfaction and catering them with expected services and products. AI becomes better with big data, which helps an organisation in understanding its customer behaviour and therefore, BI gets stronger input.

Relevant AI Information: The information provided by AI systems based on the conversion of human intelligence is of great importance, which is supported by the characteristics of comprehensiveness. The experts revealed that relevance if information helps in taking decisions on product and investments by the individual customer and also helps bank management in designing products and services as well as in taking decisions towards better customer management through better BI.

A banking or financial institution provides a number of service product lines to its customers. It has to take a number of decisions on finance, products, pricing, and promotional strategy. A wise decision can drive a bad company to a great height. On the contrary, a wrong decision may drag a firm to its demise. A great BI plays vital role in decision-making. Hence, a banking institution looks to have stronger BI that provides useful insights about inside and outside the bank to reduce the inherent risks and uncertainties involved in policy decisions.

8.4 Methodology

Based on a review of literature survey, AI enablers in the banking industry influencing BI have been identified and further validated by a total of 18 experts from banks (12 experts) and AI professionals (6 experts). Along with the finalised enablers, ISM questionnaire was prepared and administered to selected private sector banks, i.e., HDFC, ICICI Bank, Yes Bank, Indusland Bank, and Axis Bank across the top cities with digital transactions, i.e., Delhi, Bengaluru, Chennai, Coimbatore, Mumbai, Kolkata, Pune, Gurugram, Hyderabad, and Lucknow. A total of 221 questionnaires were collected and found usable for the study. The data were put through Interpretive Structure Modelling (ISM) explaining hierarchical order of enablers dependence pertaining to BI and the resultant driving and dependence variables were put forth to MICMAC analysis classifying the enablers based on their ability to influence each other revealing consistency (Figure 8.1).

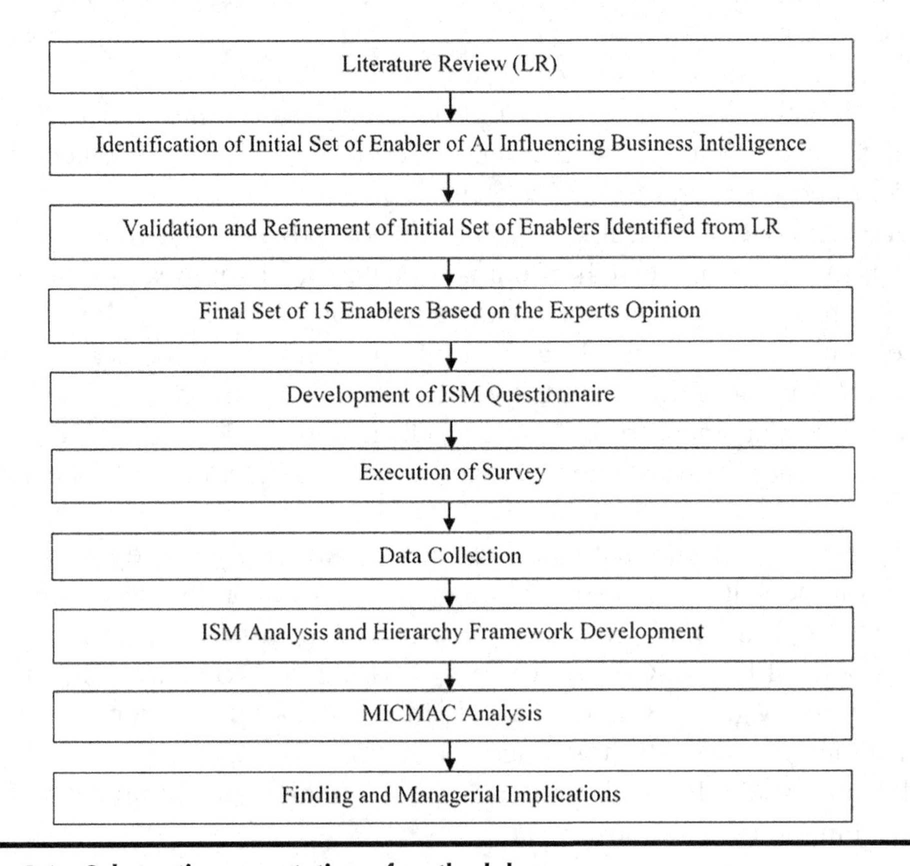

Figure 8.1 Schematic presentation of methodology.

To draw inferences on enablers and their dependence and driving power, ISM has been used to develop the enablers' hierarchy framework seeking the need for interpretation of relationships among them. Interpretive structural modelling (ISM) is a qualitative and interpretive mechanism along structural mapping of complex interconnections of elements and finds the solution by arranging the enablers or disablers in hierarchical order based on responses to ISM questionnaire. Sage, (1977) pointed out that the ISM methodology transforms unclear and poorly articulated models of systems into visible and well-defined models by decomposing the complex system into multi-level hierarchical framework development. The dependence and driving force of the enablers have been further validated by MICMAC analysis.

8.5 Results and Discussion

The adoption of AI and the implication of BI in the banking industry depends on several factors that are complex because of continuous technological innovations and upgradation. Hence, a systematic and logical approach is needed to identify and connect the elements that enable and lead to the success of AI implementation, which further strengthens BI in banking.

The model provides rational relationships among the variables (Sage, 1977). In the research, the key variables are identified using both direct and indirect interrelationships among the enablers. Figure 8.2 presents the ISM model methodology adopted in the study to reach inference and draw the hierarchy model.

Step 1: Identification of AI Enablers: The enablers have been identified based on the literature review and experts' opinion with regard to what cause the favourability to successful implication of AI, further leading to having a stronger BI in the Indian banking industry (Table 8.2).

Step 2: Development of Contextual Relation: The contextual relationship among the enablers has been reflected by banking employees/experts of selected banks and AI professional and data scientists. The relationship may be '*i* influences or causes hurdles to *j*' or '*j* influences or causes hurdles to *i*' as have been depicted in step 2 of the model where enablers along *i* have been presented column-wise and enablers along *j* have been presented row-wise.

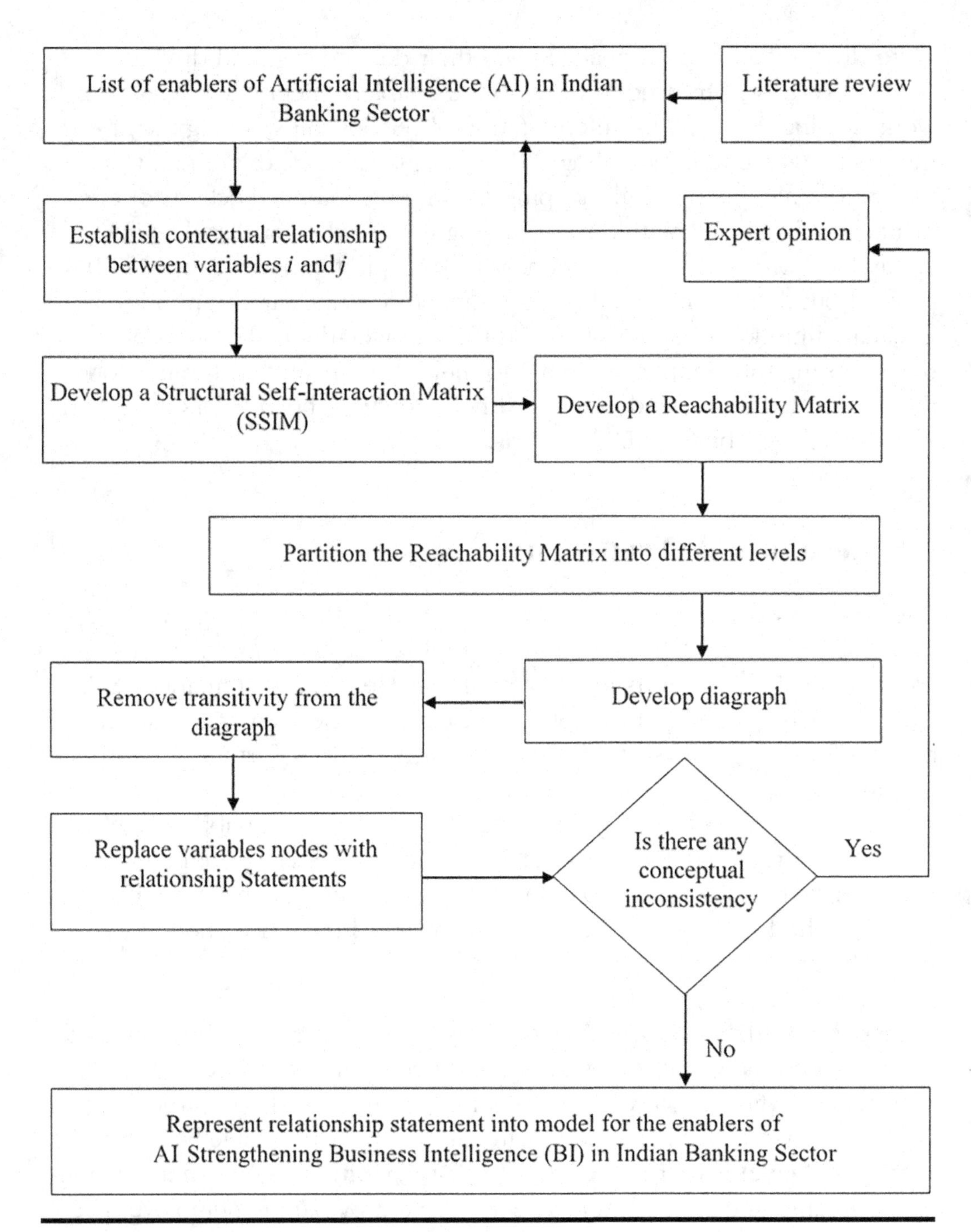

Figure 8.2 Relationship statement into model

Table 8.2 Coding of Identified Enablers

Enablers	*Code*	*Enablers*	*Code*
1. *Top Management Commitment*	EN1	9. *Multiple Query Management Capacity*	EN9
2. *Training & Skill Development*	EN2	10. *Consistent of Information*	EN10
3. *Rational Dataset*	EN3	11. *Reliability of Information*	EN11
4. *User-friendliness of AI System*	EN4	12. *Wide Use of Digital Technology*	EN12
5. *Competence to New Technology*	EN5	13. *Relevance of Information*	EN13
6. *Financial Support to Technology*	EN6	14. *Comfort of Employees with Digital Technology*	EN14
7. *Good AI System*	EN7	15. *Advanced Algorithm*	EN15
8. *Good AI Information Quality*	EN8		

Step 3: Pairwise Comparison for SSIM: In this step, the direction of relationship between the enablers has been denoted by four symbols as below:

V: Enabler *i* influences Enabler *j*;
A: Enabler *j* influences Enabler *i*;
X: Enabler *i* and *j* influence each other; and
O: Enabler *i* and *j* are unrelated (Table 8.3).

Step 4: Reachability Matrix: The SSIM is transformed into a binary matrix, called the initial reachability matrix by substituting V, A, X, O by 1 and 0 as per the case. The rules for the substitution of 1's and 0's are the following:

1. If the (*i, j*) entry in the SSIM is V, then the (*i, j*) entry in the reachability matrix becomes 1 and the (*j, i*) entry becomes 0.
2. If the (*i, j*) entry in the SSIM is A, then the (*i, j*) entry in the reachability matrix becomes 0 and the (*j, i*) entry becomes 1.
3. If the (*i, j*) entry in the SSIM is X, then the (*i, j*) entry in the reachability matrix becomes 1 and the (*j, i*) entry also becomes 1.
4. If the (*i, j*) entry in the SSIM is O, then the (*i, j*) entry in the reachability matrix becomes 0 and the (*j, i*) entry also becomes 0.

The final reachability matrix is obtained by incorporating the transitivity as enumerated in Step 4 of the ISM methodology. This is shown in

Table 8.3 Structural Self-Interaction Matrix (SSIM)

Enablers	*B15*	*B14*	*B13*	*B12*		*B11*	*B10*	*B9*	*B8*	*B7*	*B6*	*B5*	*B4*	*B3*	*B2*
EN1	V	V	0	V		0	0	V	0	V	X	V	0	V	V
EN2	0	V	0	V		0	0	V	0	0	A	V	V	V	
EN3	0	0	A	0		A	A	A	A	A	0	A	A		
EN4	A	A	0	A		0	V	X	V	V	0	A			
EN5	0	V	0	V		V	0	0	V	V	A				
EN6	V	0	0	V		0	0	V	0	V					
EN7	0	A	A	0		A	A	A	A						
EN8	A	O	A	A		X	A	A							
EN9	A	0	V	0		0	0								
EN10	0	A	X	0		X									
EN11	A	A	X	0											
EN12	X	X	V												
EN13	A	0													
EN14	X														
EN15	-														

Table 8.4. In this table, the driving power and dependence of each barrier are also shown. The driving power of a particular barrier is the total number of enablers (including it), which it may help achieve. The dependence is the total number of enablers, which may help achieving it. These driving power and dependencies will be used in the MICMAC analysis, where the enablers will be classified into four groups: autonomous, dependent, linkage, and independent (driver) enablers.

Step 5: Level Partition: Level partition in ISM is carried out based on the reachability and antecedent set (Warfield, 1974) for each enabler. The partitioning of reachability matrix has been presented in Tables 8.5–8.13.

From Table 8.5, it has been found that EN 3, i.e., Rational data, i.e., EN 3 occupies level I. Thus, it would be positioned at the top of the ISM model, and EN 1 has been found at the last level (Level 9), which is

Table 8.4 Final Reachability Matrix

Enablers	*EN1*	*EN 2*	*EN3*	*EN4*	*EN5*	*EN6*	*EN7*	*EN8*	*EN9*	*EN10*	*EN11*	*EN12*	*EN13*	*EN14*	*EN15*	*Driving Power*
EN1	1	1	0	0	1	1	1	0	1	0	0	1	0	1	1	9
EN2	0	1	1	1	1	0	0	0	1	0	0	1	0	1	0	7
EN3	0	0	1	0	0	0	0	0	0	0	0	0	0	0	0	1
EN4	0	0	1	1	0	0	1	1	1	1	0	0	0	0	0	6
EN5	0	1	0	1	1	0	1	1	0	0	1	1	0	1	0	8
EN6	1	1	0	0	1	1	1	0	1	0	0	0	1	0	1	8
EN7	0	0	1	0	0	0	1	0	0	0	0	0	0	0	0	2
EN8	0	0	1	0	0	0	1	1	0	0	1	0	0	0	0	4
EN9	0	0	1	1	0	0	1	1	1	0	0	0	1	0	0	6
EN10	0	0	1	0	0	0	1	1	0	1	1	0	1	0	0	6
EN11	0	0	1	0	0	0	1	1	0	1	1	0	1	0	0	6
EN12	0	0	0	1	0	0	0	1	0	0	0	1	1	1	1	6
EN13	0	0	1	0	0	0	1	1	0	1	1	0	1	0	0	6
EN14	0	0	0	1	0	0	1	0	0	1	1	1	0	1	1	7
EN15	0	0	0	1	0	0	0	1	1	0	1	1	1	1	1	8
Dependence power	2	4	9	7	4	2	11	9	6	5	7	6	7	6	5	

Table 8.5 Partitioning the Final Reachability Matrix (Iteration 1)

Enablers	*Reachability Set*	*Antecedent Set Intersect*	*Intersection Set*	*Level*
EN1	1,2,3,5,6,7,9,12,14,15	1	1	
EN2	2,3,4,5,9,12,14	1,2,5,6	2	
EN3	3	2,3,4,5,6,7,8,9,11,13	3	I
EN4	3,4,8,9,10	1,2,4,9,12,14,15	4,9	
EN5	2,4,5,7,8,11,12,14	1,5,6	5	
EN6	1,2,6,7,9,12,15	1,6	6	
EN7	3,7	7,10,11,13,14	7	
EN8	3,8,11	1,2,4,5,8,9,12,15	8	
EN9	3,4,8,9,13	1,4,5,9,15	4,9	
EN10	7,10,11,13	1,2,4,5,6,10,11,13,14	10,11,13	
EN11	3,7,10,11,13	1,5,8,10,11,13,14,15	10,11,13	
EN12	4,8,12,13,14,15	2,5,12,14,15	12,14,15	
EN13	3,7,10,11,13	2,6,9,10,11,12,13,15	10,11,13	
EN14	4,7,10,11,12,14,15	6,12,14,15	12,14,15	
EN15	4,8,9,11,12,13,14,15	6,12,14,15	12,14,15	

to be placed at the bottom of the diagraph and hierarchy. The identified levels help in building the digraph and the final model of ISM.

Step 6: Diagraph Development: The initial diagraph is obtained from a conical matrix including transitive links, and the final structural model is developed by removing the transitive links as shown in Figure 8.3.

Step 7: Final Structural Model: Based on the diagraph presented in Figure 8.3, final ISM hierarchy model has been constructed by the removing transitive links as shown in Figure 8.4.

The hierarchy presented in Figure 8.4 shows that top management alignment and commitment are the most significant enablers perceived by the respondents in the select banks on which training and skill development

Table 8.6 Partitioning the Final Reachability Matrix (Iteration 2)

Enablers	*Reachability Set*	*Antecedent Set Intersect*	*Intersection Set*	*Level*
EN 1	1,2,5,6,7,9,12,14,15	1	1	
EN 2	2,4,5,9,12,14	1,2,5,6	2	
EN4	4,8,9,10	1,2,4,9,12,14,15	4,9	
EN5	2,4,5,7,8,11,12,14	1,5,6	5	
EN6	1,2,6,7,9,12,15	1,6	6	
EN7	7	7,10,11,13,14	7	II
EN8	8,11	1,2,4,5,8,9,11,12,15	8	
EN9	4,8,9,13	1,4,5,9,15	4,9	
EN10	7,10,11,13	1,2,4,5,6,10,11,13,14	10,11,13	
EN11	7,10,11,13	1,5,8,10,11,13,14,15	10,11,13	
EN12	4,8,12,13,14,15	2,5,12,14,15	12,14,15	
EN13	7,10,11,13	2,6,9,10,11,12,13,15	10,11,13	
EN14	4,7,10,11,12,14,15	6,12,14,15	12,14,15	
EN15	4,8,9,11,12,13,14,15	6,12,14,15	12,14,15	

programme and financial support to technology depend. Financial support and the management commitment make the institution to adopt an advanced algorithm system, which influences the competence, comfort, and wide use of AI system. A banking institution needs to show a strong commitment towards the adoption of AI and BI for its operations and to serve its customers. The model also depicts that financial extension and training, and skill development programme is equally important, which drives all other elements towards a strong BI mechanism. The enablers placed in the upward direction based on the driving and dependence power ultimately show that strong BI is at the top in the banking system depending on all the enablers lying below. The study came up with the hierarchical order of driving and dependent variables, revealing what are the areas of attention to be paid, which would enhance the effectiveness of AI implication and, thus, that of BI in banking would help in making more rational decisions by the management.

Table 8.7 Partitioning the Final Reachability Matrix (Iteration 3)

Enablers	*Reachability Set*	*Antecedent Set Intersect*	*Intersection Set*	*Level*
EN1	1,2,5,6,9,12,14,15	1	1	
EN2	2,4,5,9,12,14	1,2,5,6	2	
EN4	4,8,9,10	1,2,4,9,12,14,15	4,9	
EN5	2,4,5,8,11,12,14	1,5,6	5	
EN6	1,2,6,9,12,15	1,6	6	
EN8	8,11	1,2,4,5,8,9,11,12,15	8,11	III
EN9	4,8,9,13	1,4,5,9,15	4,9	
EN10	8,10,11,13	1,2,4,5,6,10,11,13,14	10,11,13	
EN11	8,10,11,13	1,5,8,10,11,13,14,15	10,11,13	
EN12	4,8,12,13,14,15	2,5,12,14,15	12,14,15	
EN13	8,10,11,13	2,6,9,10,11,12,13,15	10,11,13	
EN14	4,10,11,12,14,15	6,12,14,15	12,14,15	
EN15	4,8,9,11,1,13,14,15	6,12,14,15	12,14,15	

Table 8.8 Partitioning the Final Reachability Matrix (Iteration 4)

Enablers	*Reachability Set*	*Antecedent Set Intersect*	*Intersection Set*	*Level*
EN1	1,2,5,6, 9,12,14,15	1	1	
EN2	2,4,5,9,12,14	1,2,5,6	2	
EN4	4,9,10	1,2,4,9,12,14,15	4,9	
EN5	2,4,5,7,11,12,14	1,5,6	5	
EN6	1,2,6,9,12,15	1,6	6	
EN9	4,9,13	1,4,5,9,15	4,9	
EN10	10,11,13	1,2,4,5,6,10,11,13,14	10,11,13	IV
EN11	10,11,13	1,5,8,10,11,13,14,15	10,11,13	IV
EN12	4,12,13,14,15	2,5,12,14,15	12,14,15	
EN13	10,11,13	2,6,9,10,11,12,13,15	10,11,13	IV
EN14	4,10,11,12,14,15	6,12,14,15	12,14,15	
EN15	4,9,11,12,13,14,15	6,12,14,15	12,14,15	

Table 8.9 Partitioning the Final Reachability Matrix (Iteration 5)

Enablers	*Reachability Set*	*Antecedent Set Intersect*	*Intersection Set*	*Level*
EN1	1,2,5,6,9,12,14,15	1	1	
EN2	2,4,5,9,12,14	1,2,5,6	2	
EN4	4,9	1,2,4,9,12,14,15	4,9	V
EN5	2,4,5,12,14	1,5,6	5	
EN6	1,2,6,9,12,15	1,6	6	
EN9	4,9	1,4,5,9,15	4,9	V
EN12	4,12,14,15	2,5,12,14,15	12,14,15	
EN14	4,12,14,15	6,12,14,15	12,14,15	
EN15	4,9,12,14,15	6,12,14,15	12,14,15	

Table 8.10 Partitioning the Final Reachability Matrix (Iteration 6)

Enablers	*Reachability Set*	*Antecedent Set Intersect*	*Intersection Set*	*Level*
EN1	1,2,5,6,12,14,15	1	1	
EN2	2,5,12,14	1,2,5,6	2	
EN5	2,5,12,14	1,5,6	5	
EN6	1,2,6,12,15	1,6	6	
EN12	12,14,15	2,5,12,14,15	12,14,15	VI
EN14	12,14,15	6,12,14,15	12,14,15	VI
EN15	12,14,15	6,12,14,15	12,14,15	VI

Table 8.11 Partitioning the Final Reachability Matrix (Iteration 7)

Enablers	*Reachability Set*	*Antecedent Set Intersect*	*Intersection Set*	*Level*
EN1	1,2,5,6	1	1	
EN2	2,5,6	1,2,5,6	2,5,6	VII
EN5	2,5	1,5,6	5	
EN6	1,2,6	1,6	6	

Table 8.12 Partitioning the Final Reachability Matrix (Iteration 8)

Enablers	*Reachability Set*	*Antecedent Set Intersect*	*Intersection Set*	*Level*
EN1	1,5,6	1	1	
EN5	5	1,5,6	5	VIII
EN6	1,6	1,6	1,6	VIII

Table 8.13 Partitioning the Final Reachability Matrix (Iteration 9)

Enablers	*Reachability Set*	*Antecedent Set Intersect*	*Intersection Set*	*Level*
EN1	1	1	1	IX

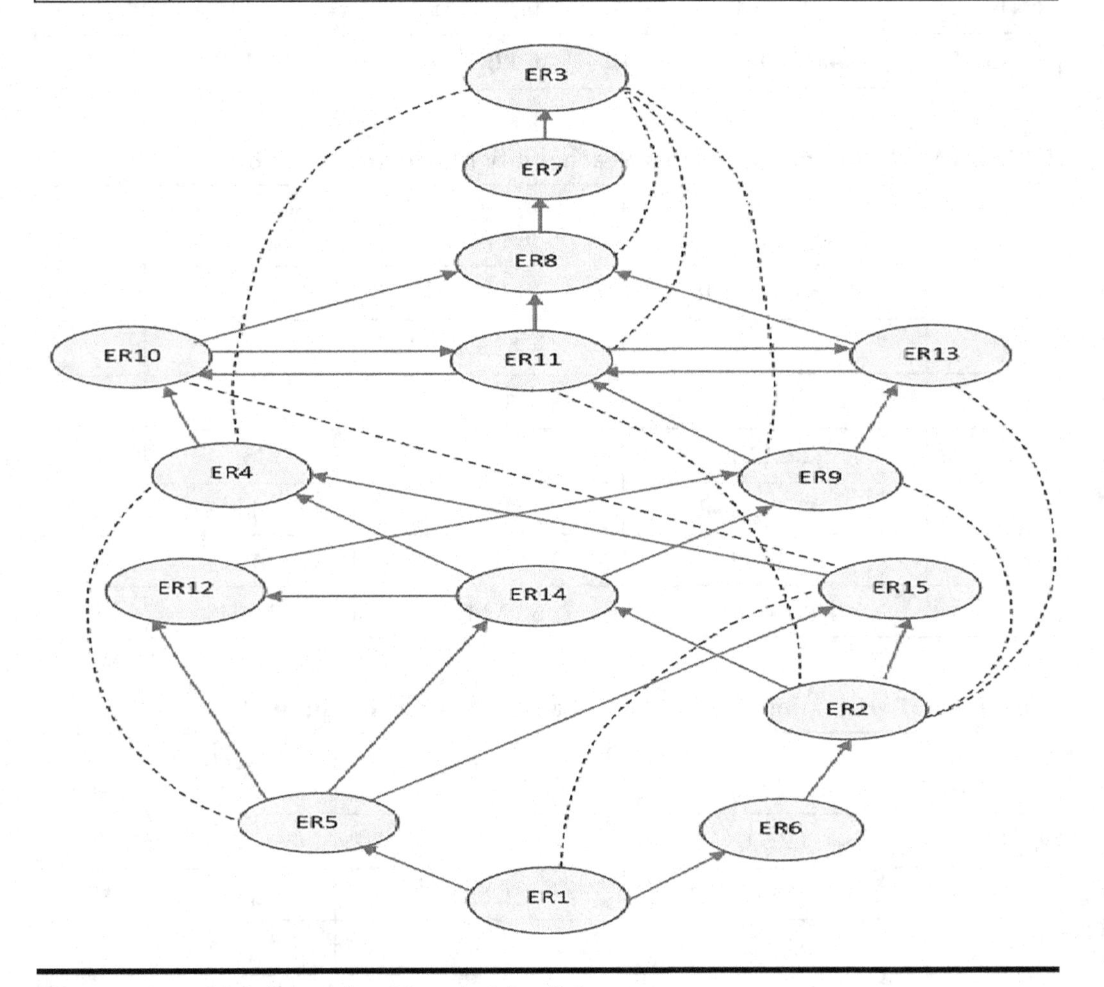

Figure 8.3 Initial diagraph with transitive link.

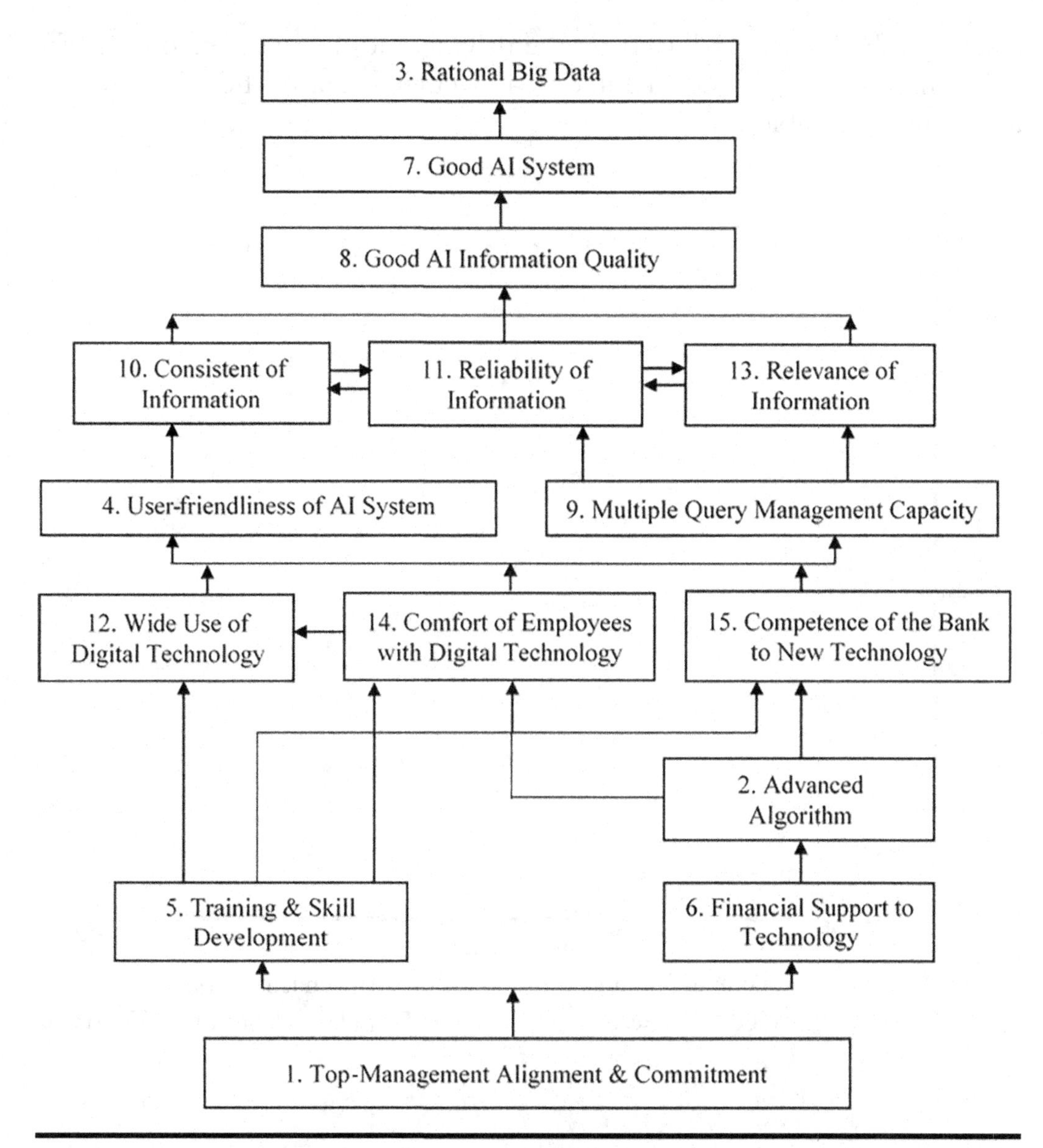

Figure 8.4 ISM model for AI enablers influencing BI in the banking industry in India.

8.6 MICMAC Analysis

The objective of the MICMAC analysis is to analyse the driver power and the dependant power of the variables. The variables are classified into four clusters and are plotted along the X-axis representing driving power and the Y-axis representing dependence in the form of matrix diagram as shown in the Figure 8.5 based on the driving and dependence power of each enabler summarised in Table 8.14 calculated out of the final reachability matrix (Table 8.14).

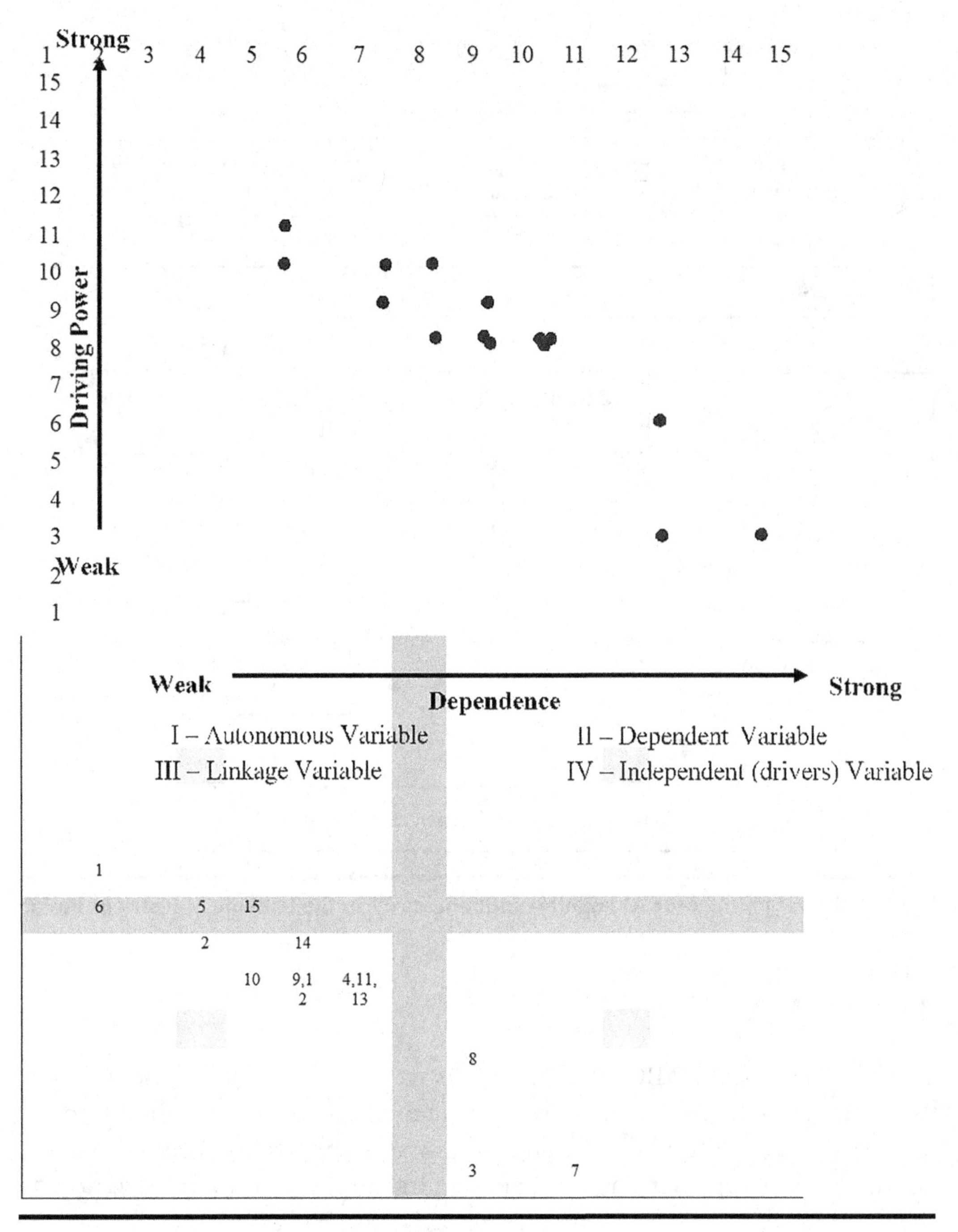

Figure 8.5 Driving power and dependence diagram (MICMAC analysis).

Table 8.14 Driving and Dependence Power of AI Enablers

Enablers	*EN1*	*EN2*	*EN3*	*EN4*	*EN5*	*EN6*	*EN7*	*EN8*	*EN9*	*EN10*	*EN11*	*EN12*	*EN13*	*EN14*	*EN15*	*Driving Power*	*Rank*
EN1	1	1	0	0	1	1	1	0	1	0	0	1	0	1	1	9	I
EN2	0	1	1	1	1	0	0	0	1	0	0	1	0	1	0	7	III
EN3	0	0	1	0	0	0	0	0	0	0	0	0	0	0	0	1	VII
EN4	0	0	1	1	0	0	1	1	1	1	0	0	0	0	0	6	IV
EN5	0	1	0	1	1	0	1	1	0	0	1	1	0	1	0	8	II
EN6	1	1	0	0	1	1	1	0	1	0	0	0	1	0	1	8	II
EN7	0	0	1	0	0	0	1	0	0	0	0	0	0	0	0	2	VI
EN8	0	0	1	0	0	0	1	1	0	0	1	0	0	0	0	4	V
EN9	0	0	1	1	0	0	1	1	1	0	0	0	1	0	0	6	IV
EN10	0	0	1	0	0	0	1	1	0	1	1	0	1	0	0	6	IV
EN11	0	0	1	0	0	0	1	1	0	1	1	0	1	0	0	6	IV
EN12	0	0	0	1	0	0	0	1	0	0	0	1	1	1	1	6	IV
EN13	0	0	1	0	0	0	1	1	0	1	1	0	1	0	0	6	IV
EN14	0	0	0	1	0	0	1	0	0	1	1	1	0	1	1	7	III
EN15	0	0	0	1	0	0	0	1	1	0	1	1	1	1	1	8	II
Dependence power	2	4	9	7	4	2	11	9	6	5	7	6	7	6	5		
Rank	VII	VI	II	III	VI	VII	I	II	IV	V	III	IV	III	IV	V		

The table depicting dependence and driving power help to conduct MICMAC analysis that clearly reveals which enablers have greater driving power and which of them are getting highly influenced. In between certain enablers fall in autonomous variable category as have been shown in Figure 8.5.

The first cluster consists of the autonomous enablers that have weak driver power and weak dependence. These enablers are relatively disconnected from the system, with which they have only a few links, which may be strong. The second cluster consists of the dependent enablers that have weak driver power but strong dependence. The third cluster has the linkage enablers that have strong driving power and strong dependence and are unstable in nature. Fourth cluster includes the independent enablers having strong driving power but weak dependence. The analysis therefore tends to be a rich source of information based on driving power and dependence tendency of the enablers, which critically defines the scope of each barrier.

The MICMAC result reflects that none of the enablers have been found in cluster III, i.e., none of the enablers are unstable. The concentration of all the enablers has been found distributed among clusters I, II, and IV. The analysis based on driving and dependence power of the enablers/drivers reveal that training and skill development pertaining to the use of AI, competence of employees to new technology, comfort of employees with the use of AI system, and financial support towards adoption of AI in the banking sector are large influencer to the BI as falls in strong driving zone, i.e., cluster IV. On the other hand, it has also been indicated that AI information quality, AI system quality, and rational data are highly influenced enablers falling in cluster II representing strong dependence zone. All the enablers are leading more or less towards rational dataset enabling the integrated BI to result in better decision-making in the banking sector.

8.7 Conclusion

Today, banking institutions are largely dependent on advanced technology and BI to make smarter decisions. The reliance has been increasing along advanced algorithms and AI implications to which data are indispensable constituents. The role of big data becomes vital today, which makes AI more effective leading to push stronger BI resulting in better decision-making by banking institutions. The integrated technology, i.e., AI and BI together is believed to serve management and customers as well by providing the requisite information and in-depth knowledge about consumer behaviour (Kautish, Singh, Polkowski, Mayura & Jeyanthi, 2021).

The ISM hierarchy model concludes that consistency, relevance, and reliability of AI information lead to quality input to both the management and customer as well. A good AI system with all the information characteristics considered tends to encourage consumers to take various decisions on investment and financial products' choice. From the employees' point of view, training and skill development, comfort of the employees to use AI frequently, top management commitment towards application of AI, and financial support are basic requirements to push a banking system towards having a smart technology. The training and skills enhance the competence of the employees in its use and drag them out of conventional thoughts. The user-friendliness of AI system's implications in banking defines its success to a greater extent creating value for the industry. It is also revealed that the quality of information and its comprehensiveness define the AI system's quality, which ultimately consequent in providing better inputs to BI and thus the firm can take better decisions on customer management and institutional strategy. The current wave of digital disruption has been unleashed by AI in the banking sector.

8.8 Implications

The study indicates several managerial implications based on the perception assessment of respondents with regard to enablers of AI implications, strengthening BI in banking. First, the AI system should be installed with an advanced algorithm, which may provide relevant and reliable information in comprehensive format. Also, the chatbots or robots used should be user-friendly and should be able to answer wider range of queries made by the customer. It will bring down the frustration of the users who may be annoyed sometimes due to limited query management of chatbots and robots. The solution will pull more users and thus there will be more interaction of customers with Ai platform. This will help AI to fetch bid data enabling system in programming human supplements into intelligence for BI to make more rational and authentic decision for customers and management.

There are still several obstructions in its adoption by all banking institutions across the spectrum of front, middle, and back office. Therefore, it can be suggested that to have an ecosystem of AI helping BI with better input, the institutions should look for incremental implementation to avoid sudden change and discomfort with operation. In any organisation, the transition should take place at acceptable pace and not overnight. It will lead to the seamless adoption and application of AI, complying with the BI and technological landscape already existing in the banking system.

8.9 Limitations

This study has several limitations. First, the perception of only employees has been measured to draw the inference with regard to AI implementation enablers' effectiveness; however, the customers also could have reflected on their experience about the quality of AI system and information. Second, the study is based on select banks, as if the responses would have been collected from all the banks in India, the perception about enablers strength and its influence on BI would have been different. Third, the population involves only the cities with the top digital transactions, although the implications of AI and BI are not limited to only the top digital transacting cities but in the banks across the country. These limitations paves way for further studies for academic community willing to conduct research in this perspective.

References

Aladwani, A.M. (2001). Change management strategies for successful ERP implementation. *Business Process Management Journal, 7*(3), 266–275.

Alwan, H.A., & Al-Zubi, A.I. (2016). Determinants of internet banking adoption among customers of commercial banks: An empirical study in the Jordanian banking sector. *International Journal of Business and Management, 11*(3), 95.

Ayyash, M.M. (2015). Identifying information quality dimensions that affect customers satisfaction of E-banking services. *Journal of Theoretical and Applied Information Technology, 82*(1), 122.

Bagale, G.S., Vandadi, V.R., Singh, D. et al. (2021). Small and medium-sized enterprises' contribution in digital technology. *Annals of Operations Research*. Doi: 10.1007/s10479-021-04235-5.

Calisira, F., & Calisirb, F. (2004). The relation of interface usability characteristics, perceived usefulness, and perceived ease of use to end-user satisfaction with enterprise resource planning (ERP) systems. *Computers in Human Behavior, 20*(4), 505–515.

Castelli, M., Manzoni, L., & Popovic, A. (2016). An artificial intelligent system to predict quality of service in banking organization. In S. Sanei (Ed.) *Computational Intelligence and Neuroscience*, 1–7. Doi: 10.1155/2016/9139380.

Delone, W.H., & McLean, E.R. (2004). Measuring e-commerce success: Applying the DeLone & McLean information system success model. *International Journal of Electronic Commerce, 9*(1), 31–47.

Elgammal, A. (2019). AI is blurring the definition of artist: Advanced algorithms are using machine learning to create art autonomously. *American Scientist, 107*(1), 18–22.

Fitriana, R., Eriyatno., & Djatna, T. (2011). Progress in Business Intelligence system research: A literature review. *International Journal of Basic & Applied Science, 11*(3), 96–105.

Goodhue, D.L., & Thompson, R.L. (1995). Task-technology fit and individual performance. *MIS Quarterly*, 213–236.

Hameed, W.U., Nadeem, S., Azeem, M., Aljumah, A.I., & Adeyemi, R.A. (2018). Determinants of e-logistic customer satisfaction: A mediating role of information and communication technology (ICT). *International Journal of Supply Chain Management, 7*(1), 105.

Huang, H.C. (2011). Technological innovation capability creation potential of open innovation: A cross-level analysis in the biotechnology industry. *Technology Analysis & Strategic Management, 23*(1), 49–63.

Hussein, R., Karim, N., Mohamed, N., & Ahlan, A. (2007). The influence of organizational factorson information systems success in E-governmentagencies in Malaysia. *The Electronic Journal of Information Systems in Developing Countries, 29*(1), 1–17.

Hwang, H.G., Ku, C.Y., Yen, D.C., & Cheng, C.C. (2004). Critical factors influencing the adoption of data warehouse technology: A study of the banking industry in Taiwan. *Decision Support Systems, 37*(1), 1–21.

Inmon, W.H., Imhoff, C., & Sousa, R. (2002). *Corporate Information Factory.* John Wiley & Sons.

Jovovic, R., Lekic, E., & Jovovic, M. (2016). Monitoring the quality of services in electronic banking. *Journal of Central Banking Theory and Practice, 5*(3), 99–119. Doi: 10.1515/jcbtp-2016-0022.

Jun, B.-H., & Kang, B.-G. (2013). Effects of information quality on customer satisfaction and continuous intention to use in social commerce. *Journal of the Korea society of computer and information,* 18(3), 127–139.

Kautish, S. (2008). Online banking: A paradigm shift. *E-Business, ICFAI Publication, Hyderabad, 9*(10), 54–59.

Kautish, S., Singh, D., Polkowski, Z., Mayura, A. & Jeyanthi, M. (2021). *Knowledge Management and Web 3.0: Next Generation Business Models.* De Gruyter, Berlin.

Kautish, S., & Thapliyal, M.P. (2013). Design of new architecture for model management systems using knowledge sharing concept. *International Journal of Computer Applications, 62*(11), 8–11.

Kumar, S. L. (2017). State of the art-intense review on artificial intelligence systems application in process planning and manufacturing. *Engineering Applications of Artificial Intelligence, 65*, 294–329.

Larivière, B., Bowen, D., Andreassen, T.W., Kunz, W., Sirianni, N.J., Voss, C., & De Keyser, A. (2017). Service Encounter 2.0: An investigation into the roles of technology, employees, and customers. *Journal of Business Research, 79*, 238–246.

Lewis, W., Agarwal, R., & Sambamurthy, V. (2003). Sources of influence on beliefs about information technology use: An empirical study of knowledge workers. *MIS Quarterly, 27*(4), 657–678.

Lin, C.C., & Chang, H.-Y.W.-F. (2011). The critical factors impact on online customer satisfaction. *Procedia Computer Science, 3*, 276–281.

Mariemuthu, C. (2019). *The Adoption of Artificial Intelligence by South African Banking Firms: A Technology, Organisation and Environment (TOE) framework*. University of the Witwatersrand, Johannesburg (*Doctoral Dissertation*).

Montargot, N., & Lahouel, B.B. (2018). The acceptance of technological change in the hospitality industry from the perspective of front-line employees. *Journal of Organizational Change Management, 31*(3), 637–655.

Moro, S., Cortez, P., & Rita, P. (2015). Business intelligence in banking: A literature analysis from 2002 to 2013 using text mining and latent dirichlet allocation. *Expert System with Applications, 42*, 1314–1324.

Munir, M.M., & Rahman, M. (2015). E-banking service quality and customer satisfaction of a state-owned schedule bank of Bangladesh. *The Journal of Internet Banking and Commerce, S2*, 009.

Najmi, M., Sepehri, M., & Hashemi, S. (2010, October). The evaluation of Business Intelligence maturity level in Iranian banking industry. *IEEE 17Th International Conference on Industrial Engineering and Engineering Management,* (pp. 466–470). IEEE.

Naumann, F., & Rolker, C. (2000). Assessment methods for information quality criteria. *14th International Conference on Information Quality,* (pp. 148–162). Potsdam, Germany.

Ngai, E. W., Hu, Y., Wong, Y. H., Chen, Y., & Sun, X. (2011). The application of data mining techniques in financial fraud detection: A classification framework and an academic review of literature. *Decision Support Systems, 50*(3), 559–569.

Rao, G. K., & Kumar, R. (2011). Framework to integrate business intelligence and knowledge management in banking industry. *Review of Business & Technology Research, 4*(1), https://arxiv.org/ftp/arxiv/papers/1109/1109.0614.pdf.

Rexha, N., Kinsgshott, R., & Shang, A. S. (2003). The inpact of the rational plan on adoption of electronic banking. *Journal of Services Marketing, 17*(1), 53–67.

Sandkuhl, K. (2019, July). Putting AI into context-method support for the introduction of Artificial Intelligence into organizations. *IEEE 21st Conference on Business Informatics (CBI), 1*, 157–164.

Seddon, P. B. (1997). A respecification and extension of the DeLone and McLean model of IS success. *Information systems research, 8*(3), 240–253.

Shu, W., & Strassmann, P. A. (2005). Does information technology provide banks with profit?. *Information & Management, 42*(5), 781–787.

Smith, W. K., & Tushman, M. L. (2005). Managing strategic contradictions: A top management model for managing innovation streams. *Organization Science, 16*(5), 522–536.

Solberg, E., Traavik, L. E., & Wong, S. I. (2020). Digital mindsets: Recognizing and leveraging individual beliefs for digital transformation. *California Management Review, 62*(4), 105–124.

Sundjaja, A. M. (2013). Implementation of business intelligence on banking, retail, and educational industry. *CommIT (Communication and Information Technology) Journal, 7*(2), 65–70.

Turban, E., Sharda, R., Aronson, J. E., & King, D. (2008). *Business Intelligence: A Managerial Approach* (pp. 58–59). Corydon, IN: Pearson Prentice Hall.

Wixom, B. H., & Todd, P. A. (2005). A theoretical integration of user satisfaction and technology acceptance. *Information Systems Research, 16*(1), 85–102. Doi: 10.1287/isre.1050.0042.

Xia, B. S., & Gong, P. (2014). Review of business intelligence through data analysis. *Benchmarking: An International Journal. 21*(2), 300–311.

Yang, Z., & Fang, X. (2004). Online service quality dimensions and their relationships with satisfaction: A content analysis of customer reviews of securities brokerage services. *International Journal of Service Industry Management, 15*(3), 302–326.

Zhang, L., Tan, J., Han, D., & Zhu, H. (2017). From machine learning to deep learning: Progress in machine intelligence for rational drug discovery. *Drug Discovery Today, 22*(11), 1680–1685.

Zohuri, B., & Rahmani, F. M. (2019). Artificial Intelligence driven resilliency with machine learning and deep learning components. *International Journal of Nanotechnology & Nanomedicine, 4*(2), 1–3.

Chapter 9

Measuring the Organizational Performance of Various Retail Formats in the Adoption of Business Intelligent

Subhodeep Mukherjee, Chittipaka Venkataiah, and Manish Mohan Baral
GITAM Institute of Management, GITAM (Deemed-to-be University)

Surya Kant Pal
Sharda University

Contents

DOI: 10.4324/9781003184928-9

9.1 Introduction

Business intelligence (BI) is defined as an approach to the recording, storage, processing, and analysis of data. BI is having five data dimensions, such as volume, variety, velocity, value, and veracity. BI will provide a creative and strategic edge for sustainable value delivery, competitive advantages, and performance assessment (Wamba et al., 2017). BI has emerged as an innovative technology in the last few years, providing many opportunities (Ramdani et al., 2013). In recent years, the demand for the implementation of BI systems has increased because of their affordability (Chaudhuri et al., 2011). However, BI implementation is not only limited to larger companies but also smaller firms. Empirical evidence shows that large companies are more likely to do so when it comes to embracing new technologies than small firms (Kannabiran, 2012). One explanation for this may be that small- and medium-sized businesses sometimes underestimate the importance of information technology (IT) (Barton & Court, 2012).

The current need is to improve retail chain management's productivity and provide cost-effective approaches for the betterment of the sector. The use of advanced market analytics, such as BI, is one opportunity (Rajesh & Saravanan, 2018). Sale figures are higher when there is a discount in comparison to other times. These figures have made the retail firms giving much value to the customers on an everyday basis in some products (Ashrafi et al., 2019). The E-commerce sector has already revolutionized the retail industry by increasing its customer reach (Nithya & Kiruthika, 2020). These give customers more options in buying a product at a comparatively lower price from online retail.

All of this would also contribute to more market competition. The e-commerce industry is one of India's critical funding areas for retail development (Rajesh & Saravanan, 2018). In modern retailing, there is also an upward trend observed. Many shopping malls are operational with various stores, such as fashion, food, and different brands in many cities (Kautish, Singh, Polkowski, Mayura & Jeyanthi, 2021). But there are many risk factors in the implementation of BI due to its high expenditure (Garg & Khurana, 2017). Here, in this research, we study the organizational performance of retail firms in the adoption process of BI in their systems. The research questions that will be addressed in this study are as follows:

RQ1: Will BI implementation in the retail sector bring changes to its organizational performance?
RQ2: Are various retail sectors ready to adopt BI in their operations?

9.2 Literature Review

9.2.1 Overview of BI

BI is utilized as an approach to settle on better choices (Kautish, 2008, Kautish & Thapliyal, 2013), fewer costs, and improve cycles and execution, having been grown principally as an instrument for addressing insightful assignments. BI, then again, has an assortment of implications. Interestingly, IBM scientist Hans Peter Luhn utilized the term BI in his article Mukherjee & Chittipaka, (2021). Insight, he said, is the capacity to consider the interrelationships of realities deciphered so that activity is coordinated toward the ideal objective (Bagale et al, 2021). This idea underscores the significance of information assortment, reports, and inquiry devices that give clients data and help them in integrating valuable information (Mukherjee et al., 2022). BI helps uncover association results, find new market openings, and make instructed business decisions about rivals, suppliers, clients, money-related issues, fundamental issues, items, and organizations (Zragat, 2020). BI applications cover most administration divisions' scientific and arranging jobs, such as showcasing, buying, and selling, monetary administration, creation of the board, the executives of advertising, control, the executives of HR, and so forth (Richards et al., 2019). In other business zones, such as hierarchical achievement, the board or client relationship, the executives, BI is likewise utilized (Seidlova et al., 2019).

9.2.2 Discussion of TOE Framework and Development of Research Hypotheses

The technological–organizational–environmental (TOE) framework defines three categories: technological, organizational, and environmental (Baker, 2012). The three segments are assessed and appear to affect the execution of innovative turns of events in technological, organizational, and environmental contexts. Concerning the technological foundation, numerous highlights will impact the appropriation of advances imperative to business, regardless of whether they are current or arising advances. The organizational foundation is portrayed by hierarchical highlights that can confine or help mechanical development in its appropriation. The environmental foundation concerns the world wherein the organization does its activities. Even though the utilization of the TOE framework is by and large upheld, it has a few downsides. What's more, that is why different investigations will utilize various factors for the TOE framework.

9.2.2.1 Technological Factors

The key technology variables are complexity, compatibility, relative advantage, and IT resources, according to many studies (Narwane, Narkhede, et al., 2019).

I. **Complexity (COMP)**: COMP was described by Rogers (1995) as the degree to which technology is viewed as challenging to comprehend and use. If they find it hard to grasp and integrate with their operational processes, enterprises are less likely to employ any technology (Verma & Bhattacharyya, 2017).
II. **Compatibility (COMPA)**: COMPA was described by Rogers (1995) as the degree to which technology is consistent with current values, past experiences, and organizational needs (Alharbi et al., 2016).
III. **Relative advantage (RADV)**: Rogers (1995) defined RADV as the benefits or advantages the technology will provide to the organizations. Compared to current technologies, if a company perceives more significant benefits from new technology, the likelihood of implementing further technology increases (Ghasemaghaei et al., 2018).
IV. **Technology assets (TA)**: Three components of TA have been identified: human assets, IT assets, and related assets (Magaireah et al., 2017). The employee's technological skills to comprehend, interpret, and grasp the market domain are human assets. IT assets are the tools and

technologies needed for technology adoption, such as hardware, software, frameworks, and databases (Puklavec et al., 2018).

Hence, H1: TF will positively impact BIA for OP.

9.2.2.2 Organizational Factors

OF includes factors that affect the firms' organizational prescriptive, including support from top management, cost, and investment (Narwane, Narkhede et al., 2019).

I. **Top Management Support (TMS)**: TMS in prior research is generally supported for innovation adoption. TMS ensures the successful adoption of BI through the development of the BI plan and vision, the funding of BI projects, and the danger of BI project implementation (Raut et al., 2019).
II. **Organization Data Availability (ODA)**: The general characterized information rules and guidelines for information assortment, stockpiling, and investigation are fundamental parts of establishing a steady hierarchical information climate for selecting BI ODA is the information availability of the organizations. The very much characterized information rules and guidelines for information assortment, stockpiling, and investigation are fundamental parts of establishing a steady, authoritative information climate for selecting BI (Kumar & Krishnamoorthy, 2020).
III. **Perceived Costs (PC)**: PC is estimated as capital consumption and working costs. Startup costs, operating expenses, preparing costs, and managerial costs identified with the administration and usage are remembered for the monetary costs.
IV. **Financial Investment Competence (FIC)**: FIC refers to how many organizations can put money into the presentation and service of BI (Narwane, Raut et al., 2019). It takes a great deal of monetary undertaking to procure BI in companies, including facilities, programming packages, and consultation (Garg & Khurana, 2017).

Hence, H2: OF will positively impact BIA for OP.

9.2.2.3 Environmental Factors

Technology adoption studies have predominantly established competitive pressure (CP) and market pressure as key environmental factors (EFs) (Rajesh & Saravanan, 2018).

I. **CP**: CP impacts the severe climate of the organization by utilizing innovation to keep up and increment intensity (Shahzad et al., 2020). Organizations with more rivalry for the overall industry, income, market advancement, and item creation efficiency are bound to utilize business investigation.
II. **Industry Pressure (IP)**: The industry's burden is the impact of the kind of industry technology adoption that belongs to the company. The type of industry that a company belongs to determines the adoption of technology in an organization (Dlima et al., 2020).
III. **External Support (ES)**: It has been described as one of IT's key drivers. Achievement of growth can emphatically affect IT reception of progression (Nam et al., 2020).

Hence, H3: EF will positively impact BIA for OP.

9.2.2.4 BI Adoption

Technology implementation appears to positively influence the company's efficiency (Richards et al., 2019). BI will help firms in reviving their operational excellence and increase customer satisfaction.

Hence, H4: BIA will positively help the firms for O.P.

9.3 Research Methodology

9.3.1 Sampling

Responses were gathered through an organized survey from the retail supervisors, activity administrators, departmental staff, and the head supervisor of the different retail locations. Simple random sampling is being utilized for the assortment of information (Mukherjee, Mohan Baral et al., 2021). The survey questionnaire was sent to 459 respondents, yet just 271 returned the questionnaires, legitimate for investigation. To stay away from a typical technique, predisposition, the examination group has avoided potential risk during the pre-information assortment stage (Mukherjee, Chittipaka et al., 2021). At the start of the survey, a note showed the study was planned for scholarly examination, and information secrecy would be kept up with. In the assembled dataset, the main purifying was done by case screening, followed by factor screening with the goal that an explanation could be given to the assortment in the data (Baral & Verma, 2021). It is a need to follow this cycle

Table 9.1 Demographics of the Respondents

Sl. No.	*Characteristics*	*Percentage*
1	**Gender**	
1.A	Male	57
1.B	Female	43
2	**Respondent's current position**	
2.A	Retail manager	25
2.B	Operation manager	27
2.C	Departmental manager	19
2.D	Store manager	29

so that there were no missing characteristics in the dataset. Be that as it may, after the information is gathered, the exploration group applied Harman's single factor test. An exploratory factor examination was performed, and the outcomes show that the primary factor clarifies the most extreme change (33.838%) underneath the suggested worth of half (Podsakoff, 2003).

9.3.2 Demographics of the Respondents

Leedy and Ormrod (2014) stated that a cross-sectional plan includes testing and looking at individuals from a few diverse segment gatherings. This methodology empowers the specialist to gather the essential information simultaneously. Table 9.1 summarizes the demographics of the respondents.

9.4 Data Analysis

9.4.1 Reliability and Validity

9.4.1.1 Cronbach's Alpha

The reliability test was performed for each factor dependent on Cronbach's alpha (α) esteem presents Cronbach's alpha for the development. The significance, all things considered, or dimensional scales ought to be over the suggested worth of 0.70 (Mueller & Hancock, 2018). Use of the seven-point Likert scale was done in setting up the organized survey. Subsequently, the results of the reliability test are displayed in Table 9.2.

Table 9.2 Cronbach's Alpha, Composite reliability, Rotated Component Matrix, AVE for the Variables

Latent Variable	*Indicators*	*Cronbach's Alpha (α)*	*Rotated Component Matrix*	*AVE*
TF	COMP	0.854	0.800	0.538
	COMPA		0.863	
	RADV		0.895	
	T.A.		0.788	
OF	TMS	0.886	0.864	0.537
	ODA		0.835	
	PC		0.916	
	FIC		0.838	
EF	CP	0.849	0.892	0.506
	IP		0.923	
	ES		0.809	

9.4.2 Exploratory Factor Analysis

The initial step of the EFA was to assess the propriety of the example size. SPSS 20.0 was used for EFA. The connections between its things had been examined utilizing Bartlett's trial of sphericity (Lefever et al., 2007). The principal axis factoring was performed to distinguish significant inclination and express similar characteristics. The KMO value for the research was 0.700. The base-level set for this measurement was 0.60 (Keith et al., 1995). The importance esteem was 0.000, which was under 0.05, i.e., the likelihood esteem level was satisfactory. The eigenvalues which have values more prominent than one were separated as it clarifies the greatest difference. The Rotated Component Matrix is fundamental for deciphering the consequences of the investigation. Turn helps bunch the things, and each gathering contains multiple things at any rate, which works on the construction (Netemeyer et al., 2003). Thus, this is the point of the objective of pivot. In this examination, we have accomplished this point. There are 11 all out factors, which were assembled under three unique segments, as displayed in Table 9.2.

Table 9.3 Construct Correlation and AVE

	CR	*AVE*	*MSV*	*MaXR (H)*	*TP*	*OP*	*EP*
TF	0.849	0.538	0.181	0.883	0.733		
OF	0.840	0.537	0.116	0.906	0.340	0.733	
EF	0.783	0.506	0.181	0.856	0.426	0.268	0.712

9.4.3 Construct Validity

It's a good idea to test the validity of a construct validity (CV) measure. When a test quantifies an idea or development, it's called CV. CV is usually tested by estimating the relationship between a few scales of evaluations. CV does not have a cut-off point (Sánchez et al., 2005) (Table 9.3).

9.4.4 Structural Equation Modeling

The latent variables along with their indicators are technological factors (TFs): TFs have four indicators: COMP, COMPA, RADV, and TA; OF: Organizational factors have four indicators TMS, ODA, PC, and FIC; EF: EFs have three indicators: CP, IP, and ES. One mediating variable is BIA: Business Intelligence Adoption, which has four indicators: BIA1, BIA2, BIA3, and BIA4. One dependent variable is OP: Organizational Performance, which has four indicators OP1, OP2, OP3, and OP4. Table 9.4 summarizes the values of model fit parameters (Figure 9.1).

Table 9.4 summarizes the results of the path estimate analysis. Using the P-value, the four hypotheses are confirmed (Byrne, 2010; Hair et al., 2012). If you want to find out how well a regression line predicts real data points between 0 and 1, you can look at the multiple square correlations (R2) (Hair et al., 2014). The model's ability to predict that technology improves as the value gets closer to 1. According to the proposed model, OP variance can be explained by 54.9% (Table 9.5).

9.5 Discussion

BIA in retail sectors will help the industry achieve operational excellence (Eder & Koch, 2018). The most significant factor in the successful implementation of BI is the customization of the BI system according to the

Table 9.4 Model Fit Parameters

Goodness-of-Fit Indices	*Default Model*	*Benchmark*
Chi-square	612.316	
Degrees of freedom	145	
Probability level	0.000	
Absolute Goodness-of-Fit Measure		
χ^2/df (CMIN/DF)	4.223	Lower limit:1.0 upper limit 2.0/3.0 or 5.0
GFI	0.830	>0.80
RMSEA	0.06	<0.08
Incremental Fit Measure		
CFI	0.958	⩾0.80
IFI	0.859	⩾0.80
TLI	0.833	⩾0.80

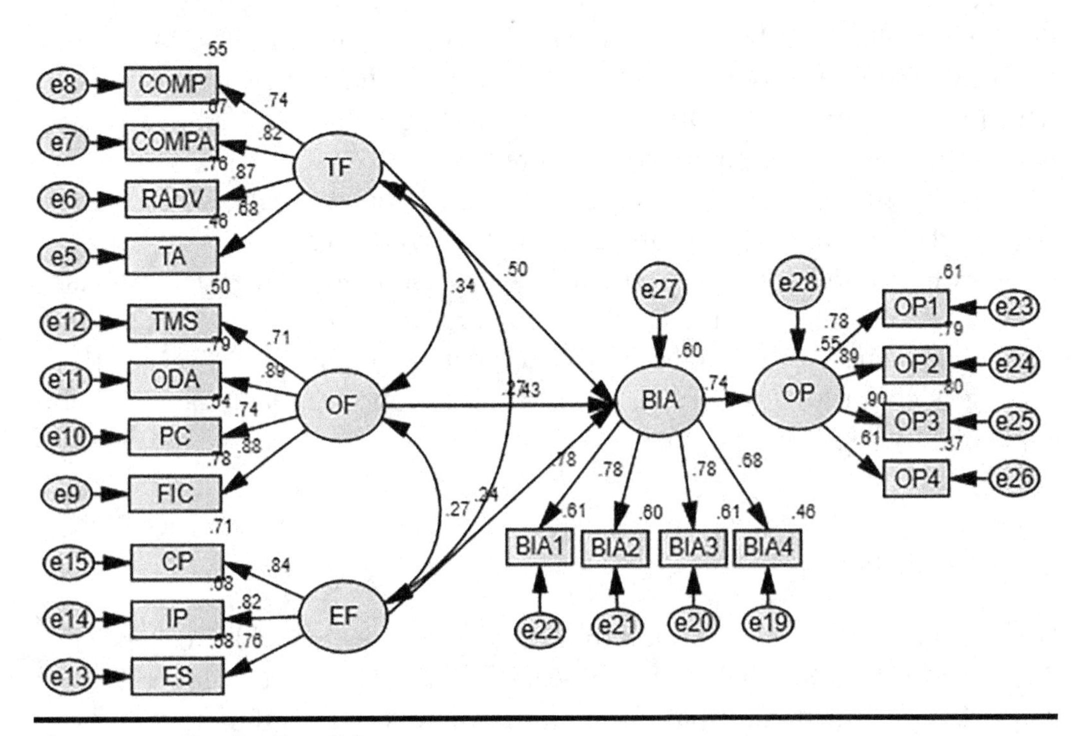

Figure 9.1 The final model.

Table 9.5 Structural Model Results

	Estimate	*SE*	*CR*	*P*	*Hypothesis*
BIA <— EF	0.244	0.049	4.979	0.000	Supported
BIA <— OF	0.266	0.046	5.783	0.000	Supported
BIA <— TF	0.496	0.083	5.976	0.000	Supported
OP <— BIA	0.741	0.129	5.744	0.000	Supported

requirements of managers (Apraxine & Stylianou, 2017). The indicators that had a significant impact are COMP, COMPA, RADV, TA, TMS, ODA, PC, FIC, CP, IP, and ES. The component TF relates to the technological aspects for adopting BIA. It comprises four sub-components: COMP, COMPA, RADV, and TA, and each loading is 0.828, 0.863, 0.885, and 0.771. All the organizations agreed that IT asset availability is crucial to BA's adoption (Acheampong et al., 2017). Many prior studies have also supported IT infrastructure as an essential factor in adopting technology (Raut et al., 2019; Richards et al., 2019).

OF relates to organizational aspects for adopting BIA. It comprises four sub-components: TMS, ODA, PC, and FIC and each loadings are 0.856, 0.863, 0.882, and 0.858. The main determinant of recent technology adoption is the encouragement from top management support (Seidlova et al., 2019; Shanmugam et al., 2020). Employees see top management as spending financial capital for the business's benefit through investment and implementation of BI. From an employee's point of view, TMS often suggests that the organization can take risks and is interested in achieving a competitive advantage (Sianipar et al., 2019). Data sourcing, storage regulations, usability, data consistency, and a data-driven culture have all been recognized as important factors in organizations' data management environments. According to many studies on cloud computing adoption, perceived cost is a major factor (Puklavec et al., 2018; Ramdani et al., 2013). Rather than becoming a facilitator of market analytics, the estimated expense acts as an obstacle (Yiu, Yeung, & Cheng, 2020; Yiu, Yeung, & Jong, 2020). EF relates to the environmental aspects of adopting BIA. It comprises three sub-components: CP, IP, and ES, and each loading is 0.888, 0.897, and 0.843. Hence, the loadings of sub-components are >|.40|. As one determinant of new technology's adoption, CP complies with most research (Zragat, 2020). Several organizations said they closely track the technical advances implemented by rival companies (Ghasemaghaei et al., 2018).

Construct validity is also an essential component of the analysis. Hence, AVE was calculated, which is >0.5 for all three constructs, TF, OF, and EF, which satisfies the convergent validity for all the constructs. Further divergent or discriminant validity was also checked for all the three constructs, which shows MSV<AVE. Hence, this criterion was also satisfied.

9.6 Conclusion

This study aims to measure the organizational performance in adopting BI in various retail sectors. BI provides many benefits to the industry and helps them increase operational excellence, which helps gain more profits. This study was conducted by searching and identifying the factors that impact BI. We adopted TOE frameworks for this study as much of the latest technological innovative research is being done. This framework helps us understand three critical elements, such as technological, organizational, and EFs. There is currently work being done on a questionnaire for a survey-based study on retail. People working in these fields were primarily targeted. A preliminary factor analysis and structural equation modeling are conducted after the data has been collected and analyzed. The developed model fit the data well, and all of the study's hypotheses were confirmed.

References

Acheampong, O., Agbemabiese, G. C., & Soladoye, A. (2017). Determinants of business intelligence systems adoption in developing countries. *Journal of Internet Banking and Commerce, 22*(S8), 1–15. http://www.icommercecentral.com.

Alharbi, F., Atkins, A., & Stanier, C. (2016). Understanding the determinants of cloud computing adoption in Saudi healthcare organisations. *Complex & Intelligent Systems, 2*(3), 155–171. Doi: 10.1007/s40747-016-0021-9.

Apraxine, D., & Stylianou, E. (2017). Business intelligence in a higher educational institution. *Proceedings of 2017 IEEE Global Engineering Education Conference (Educon2017),* April, 1735–1746.

Ashrafi, A., Zare Ravasan, A., Trkman, P., & Afshari, S. (2019). The role of business analytics capabilities in bolstering firms' agility and performance. *International Journal of Information Management, 47*(December 2018), 1–15. Doi: 10.1016/j.ijinfomgt.2018.12.005.

Bagale, G. S., Vandadi, V. R., Singh, D. et al. (2021). Small and medium-sized enterprises' contribution in digital technology. *Annals of Operations Research.* Doi: 10.1007/s10479-021-04235-5.

Baker, J. (2012). *The Technology–Organization–Environment Framework* (pp. 231–245). Springer, New York. Doi: 10.1007/978-1-4419-6108-2_12.

Baral, M. M., & Verma, A. (2021). Cloud computing adoption for healthcare: An empirical study using SEM approach. *FIIB Business Review*, 231971452110125. Doi: 10.1177/23197145211012505.

Barton, D., & Court, D. (2012). Making advanced analytics work for you. *Harvard Business Review*, 90(10), 78–83.

Byrne, B. M. (2010). Structural equation modeling with AMOS: basic concepts, applications, and programming (multivariate applications series). New York: Taylor & Francis Group, 396(1), 7384.

Chaudhuri, S., Dayal, U., & Narasayya, V. (2011). An overview of business intelligence technology. *Communications of the ACM*, *54*(8), 88–98. Doi: 10.1145/1978542.1978562.

Dlima, C., Kamath, D. B., Kumar, A., & Krishnamoorthy, B. (2020). Business analytics adoption in firms. *International Journal of Business Information Systems*, *1*(1), 1. Doi: 10.1504/ijbis.2020.10030813.

Eder, F., & Koch, S. (2018). Critical success factors for the implementation of business intelligence systems. *International Journal of Business Intelligence Research*, *9*(2), 27–46. Doi: 10.4018/IJBIR.2018070102.

Garg, P., & Khurana, R. (2017). Applying structural equation model to study the critical risks in ERP implementation in Indian retail. *Benchmarking*, *24*(1), 143–162. Doi: 10.1108/BIJ-12-2015-0122.

Ghasemaghaei, M., Ebrahimi, S., & Hassanein, K. (2018). Data analytics competency for improving firm decision making performance. *Journal of Strategic Information Systems*, *27*(1), 101–113. Doi: 10.1016/j.jsis.2017.10.001.

Hair, J. F., Sarstedt, M., Hopkins, L., & Kuppelwieser, V. G. (2014). Partial least squares structural equation modeling (PLS-SEM): An emerging tool in business research. *European Business Review*, *26*(2), 106–121. Emerald Group Publishing Ltd. Doi: 10.1108/EBR-10-2013-0128.

Hair, J. F., Sarstedt, M., Ringle, C. M., & Mena, J. A. (2012). An assessment of the use of partial least squares structural equation modeling in marketing research. *Journal of the Academy of Marketing Science*, *40*(3), 414–433. Doi: 10.1007/s11747-011-0261-6.

Kautish, S. (2008). Online banking: A paradigm shift. *E-Business, ICFAI Publication, Hyderabad*, *9*(10), 54–59.

Kautish, S., Singh, D., Polkowski, Z., Mayura, A., & Jeyanthi, M. (2021). *Knowledge Management and Web 3.0: Next Generation Business Models*. De Gruyter, Berlin.

Kautish, S., & Thapliyal, M.P. (2013). Design of new architecture for model management systems using knowledge sharing concept. *International Journal of Computer Applications*, *62*(11).

Kannabiran, G. (2012). Enablers and inhibitors of advanced information technologies adoption by SMEs: An empirical study of auto ancillaries in India. *Journal of Enterprise Information Management*, *25*(2), 186–209. Doi: 10.1108/17410391211204419.

Keith, T. Z., Fugate, M. H., Degraff, M., Diamond, C. M., Shadrach, E. A., & Stevens, M. L. (1995). Using multi-sample confirmatory factor analysis to test for construct bias: An example using the K-ABC. *Journal of Psychoeducational Assessment, 13*(4), 347–364. Doi: 10.1177/073428299501300402.

Kumar, A. & Krishnamoorthy, B. (2020). Business analytics adoption in firms: A qualitative study elaborating TOE framework in India. *International Journal of Global Business and Competitiveness*, 15(2), 80–93.

Leedy, P.D. & Ormrod, J. E. (2014). *Practical Research, Planning and Design* (10th ed.). Pearson Education Inc., Upper Saddle River, NJ.

Lefever, S., Dal, M., & Matthíasdóttir, Á. (2007). Online data collection in academic research: Advantages and limitations. *British Journal of Educational Technology, 38*(4), 574–582. Doi: 10.1111/J.1467-8535.2006.00638.X.

Magaireah, A. I., Sulaiman, H., & Ali, N. (2017). Theoretical framework of critical success factors (CSFs) for Business Intelligence (BI) System. *ICIT 2017–8th International Conference on Information Technology, Proceedings*, 455–463. Doi: 10.1109/ICITECH.2017.8080042.

Mueller, R. O., & Hancock, G. R. (2018). *Structural Equation Modeling*. Routledge.

Mukherjee, S. & Chittipaka, V. (2021). Analysing the adoption of intelligent agent technology in food supply chain management: an empirical evidence. *FIIB Business Review*, 23197145211059243.

Mukherjee, S., Chittipaka, V., & Baral, M. M. (2021). Developing a model to highlight the relation of digital trust with privacy and security for the blockchain technology. In *igi-global.com*, 110–125. Doi: 10.4018/978-1-7998-8081-3.ch007.

Mukherjee, S., Chittipaka, V., & Baral, M. M. (2022). Addressing and modeling the challenges faced in the implementation of blockchain technology in the food and agriculture supply chain: A study using TOE framework. In *Blockchain Technologies and Applications for Digital Governance* (pp. 151–179). IGI Global.

Mukherjee, S., Mohan Baral, M., Srivastava, S. C., & Jana, B. (2021). Analyzing the problems faced by fashion retail stores due to covid-19 outbreak. *Parikalpana-KIIT Journal of Management, 17*(I). Doi: 10.23862/kiit-parikalpana/2021/v17/i1/209031.

Nam, K., Dutt, C. S., Chathoth, P., Daghfous, A., & Khan, M. S. (2020). The adoption of artificial intelligence and robotics in the hotel industry: Prospects and challenges. *Electronic Markets*. Doi: 10.1007/s12525-020-00442-3.

Narwane, V. S., Narkhede, B. E., Gardas, B. B., & Raut, R. D. (2019). Cloud manufacturing issues and its adoption: Past, present, and future. *International Journal of Management Concepts and Philosophy, 12*(2), 168. Doi: 10.1504/ijmcp.2019.099319.

Narwane, V. S., Raut, R. D., Gardas, B. B., Kavre, M. S., & Narkhede, B. E. (2019). Factors affecting the adoption of cloud of things: The case study of Indian small and medium enterprises. *Journal of Systems and Information Technology, 21*(4), 397–418. Doi: 10.1108/JSIT-10-2018-0137.

Netemeyer, R., Bearden, W., & Sharma, S. (2003). *Scaling Procedures: Issues and Applications*. https://books.google.com/books?hl=en&lr=&id=woiECgAAQBAJ&oi=fnd&pg=PR11&dq=Netemeyer,+2003&ots=MC5yok9s8N&sig=U_Odqt2MPflrduqvJAmsPU5Punc.

Nithya, N., & Kiruthika, R. (2020). Impact of Business Intelligence adoption on performance of banks: A conceptual framework. *Journal of Ambient Intelligence and Humanized Computing, 0123456789*. Doi: 10.1007/s12652-020-02473-2.

Podsakoff, N. P. (2003). Common method biases in behavioral research: A critical review of the literature and recommended remedies. *Journal of Applied Psychology, 885*(879), 10–1037.

Puklavec, B., Oliveira, T., & Popovič, A. (2018). Understanding the determinants of business intelligence system adoption stages an empirical study of SMEs. *Industrial Management and Data Systems, 118*(1), 236–261. Doi: 10.1108/IMDS-05-2017-0170.

Rajesh, K., & Saravanan, D. (2018). Applying structural equation model to study the critical risks in business intelligence and analytical system implementation in Indian retail. *International Journal of Management Concepts and Philosophy, 11*(2), 190. Doi: 10.1504/ijmcp.2018.092338.

Ramdani, B., Chevers, D., & Williams, D. A. (2013). SMEs' adoption of enterprise applications: A technology-organisation-environment model. *Journal of Small Business and Enterprise Development, 20*(4), 735–753. Doi: 10.1108/JSBED-12-2011-0035.

Raut, R. D., Gardas, B. B., Narkhede, B. E., & Narwane, V. S. (2019). To investigate the determinants of cloud computing adoption in the manufacturing micro, small and medium enterprises: A DEMATEL-based approach. *Benchmarking, 26*(3), 990–1019. Doi: 10.1108/BIJ-03-2018-0060.

Richards, G., Yeoh, W., Chong, A. Y. L., & Popovič, A. (2019). Business Intelligence effectiveness and corporate performance management: An empirical analysis. *Journal of Computer Information Systems, 59*(2), 188–196. Doi: 10.1080/08874417.2017.1334244.

Rogers, E. M. (1995). Diffusion of innovations: Modifications of a model for telecommunications. In *Die Diffusion von Innovationen in der Telekommunikation* (pp. 25–38). Springer Berlin Heidelberg. Doi: 10.1007/978-3-642-79868-9_2.

Sánchez, B. N., Budtz-JØrgensen, E., Ryan, L. M., & Hu, H. (2005). Structural equation models: A review with applications to environmental epidemiology. *Journal of the American Statistical Association, 100*(472), 1443–1455. Taylor & Francis. Doi: 10.1198/016214505000001005.

Seidlova, R., Poživil, J., & Seidl, J. (2019). Marketing and business intelligence with help of ant colony algorithm. *Journal of Strategic Marketing, 27*(5), 451–463. Doi: 10.1080/0965254X.2018.1430058.

Shahzad, F., Xiu, G. Y., Khan, I., Shahbaz, M., Riaz, M. U., & Abbas, A. (2020). The moderating role of intrinsic motivation in cloud computing adoption in online education in a developing country: A structural equation model. *Asia Pacific Education Review, 21*(1), 121–141. Doi: 10.1007/s12564-019-09611-2.

Shanmugam, K., Jeganathan, K., Mohamed Basheer, M. S., Mohamed Firthows, M. A., & Jayakody, A. (2020). Impact of business intelligence on business performance of food delivery platforms in Sri Lanka. *Global Journal of Management and Business Research, 20*(6), 39–51. Doi: 10.34257/gjmbrgvol20is6pg39.

Sianipar, K. C., Wicaksana, S., Parikenan, B., & Hidayanto, A. N. (2019). Business Intelligence critical success factors evaluation using analytical hierarchy process. *5th International Conference on Computing Engineering and Design, ICCED 2019*, 1–6. Doi: 10.1109/ICCED46541.2019.9161108.

Verma, S., & Bhattacharyya, S. S. (2017). Perceived strategic value-based adoption of Big Data analytics in emerging economy: A qualitative approach for Indian firms. *Journal of Enterprise Information Management, 30*(3), 354–382. Doi: 10.1108/JEIM-10-2015-0099.

Wamba, S. F., Gunasekaran, A., Akter, S., Ren, S. J. fan, Dubey, R., & Childe, S. J. (2017). Big data analytics and firm performance: Effects of dynamic capabilities. *Journal of Business Research, 70*, 356–365. Doi: 10.1016/j.jbusres.2016.08.009.

Yiu, L. M. D., Yeung, A. C. L., & Cheng, T. C. E. (2020). The impact of business intelligence systems on profitability and risks of firms. *International Journal of Production Research, 0*(0), 1–24. Doi: 10.1080/00207543.2020.1756506.

Yiu, L. M. D., Yeung, A. C. L., & Jong, A. P. L. (2020). Business intelligence systems and operational capability: An empirical analysis of high-tech sectors. *Industrial Management and Data Systems, 120*(6), 1195–1215. Doi: 10.1108/IMDS-12-2019-0659.

Zragat, O. M. (2020). The moderating role of business intelligence in the impact of big data on financial reports quality in Jordanian telecom companies. *Modern Applied Science, 14*(2), 71. Doi: 10.5539/mas.v14n2p71.

Chapter 10

Implementation and Scope of Business Intelligence and Oracle Transportation Management System in Tata Steel Supply Chain

Abhinav Verma and Nitesh Kumar Adichwal
University of Petroleum and Energy Studies (UPES)

Prayas Sharma
Indian Institute of Management, Sirmaur (IIM)

Contents

DOI: 10.4324/9781003184928-10

10.1 Introduction

The Tata Group was founded in Jamshedpur (India) in 1907 under the vision of founder Jamsetji N. Tata. In today's market, the Tata Group is regarded as the world's most highly dispersed steel-producing company due to the company's economic and trade presence in the global market.[1] Around 65,000 employees work for the company, which operates in five continents. The company is committed to and concentrated on Technology, Innovation, People, and Sustainability, and strives to maintain its position as the global steel manufacturing benchmark for value development, as well as to become the most trusted and admired brand in the metals and minerals sector. Tata Steel currently produces 19.6 million tons of crude steel in India through manufacturing facilities in Dhenkanal, Kalinganagar, Jharkhand, Sahibabad, Uttar Pradesh, and Khopoli, Maharashtra. Tata Steel has also begun the first phase of its two-phase expansion plan for its Kalinganagar steel plant, which will increase capacity to 8 million tons per annum.

As a result, the company has several downstream product extensions, including manufacturing facilities for Bearings, Agriculture, Wires, Tubes, and Equipment. Additionally, the company operates a ferroalloys and minerals division, as well as a heavy-duty fabrication and engineering division (Kautish, 2008, Kautish and Thapliyal, 2013). Tata Steel was able to manufacture 16.26 million tons of steel per annum for the Indian market in FY 2019, a 34% increase over FY 2018. This increase was facilitated by the acquisition of Bhushan Steel (TSBSL), which benefited both Kalinganagar and Tata Steel BSL. Tata Steel currently operates as an end-to-end value chain, beginning with mining and ending with finished steel goods, serving a diverse range of market segments, including general, engineering, and automotive.

The company owns and operates captive mines, which enables it to maintain cost-competitiveness and production efficiency by ensuring the raw material supply chain remains uninterrupted (Singh & Gite, 2015). Tata Steel's

[1] https://www.tata.com/business/tata-steel accessed at 4PM on 12/05/2021.

iron ore mines in Jharkhand and Odisha ensure complete captive iron ore utilization, while its coal mines in West Bokaro and Jhanria ensure 30% captive coal utilization. Tata Steel owns one dolomite mine and one chromate mine, allowing for a consistent and efficient supply of raw materials to its ferro alloy plant. Robust efficiency in the supply chain of a business enables it to achieve greater economic efficiency and customer satisfaction standards.

In the international market of Europe, Tata Steel is one of the largest steel producers with a capacity of 12.3 Mn TPA in crude steel production. It was able due to its acquisition of Corus in 2007, which helped them to make its presence in the market there. Tata Steel has steel-making facilities in the United Kingdom and the Netherlands and plants in Europe; this helps in the movement of the strip steel products in the supply chain of the company to fulfill the demand of the market, such as construction, automotives, packaging, etc.

In the countries of South-East Asia, Tata Steel started its operations in 2004 with the acquisition of NatSteel, Singapore. In 2005, Tata also took a majority stake in Thailand-based steelmaker Millennium Steel, which helped the company to strengthen its market.

10.2 Literature Review

Today's market is very dynamic for businesses as it can change at any time and at any point, but this technology adaptation was also a great acceptance for all businesses. The same was with the supply chain to move from traditional to the new emerging world (Bagale et al., 2021). As Rejeb et al. (2018) and Smock et al. (2007) mentioned in their articles about the change in the traditional form of one to one between the buyer and the seller, but with the development of the technology, there can be seen a huge movement of technology as it can be accessed and visible with the supply chain partners. Tipping and Kauschke (2016), Cheung and Bal (1998), and Davenport and Short (1990) in their articles also mentioned many technologies that are being already used around the world, but the implementation of those technologies is still a big part from automation to the block chain movement of the data openly.

With the employment of amazing new regulating technology for the flow of goods along the supply chain, the technology not only increases the number of workers, but also reduces the business's risk exposure. In his 2017 essay for Deloitte, Brian Umbenhauer discusses "Source to Contract," which uses artificial intelligence to analyse costs and determine the optimal contract strategy for cost reduction. In addition, he discusses how business development for relationship management using "Supplier Relationship Management" aids in understanding the dangers that might be produced.

The movement in technology is not only one part of the movement; it is a whole movement of the economy with the sectors to develop an overall infrastructure so that every part and process can develop and adapt itself with the help of re-engineering of technology according to the requirement of the process. Groznik and Maslaric (2010) in their research talked about the movement of the fundamental tools with the development of business technology-driven as data and was considered important for business, but today with technology it is also creating and moving for factual decision-making. In the early 1990s, re-engineering was implemented to capitalise on the arrival of information technology, but with time it evolved into an integral element of the business. ERP is becoming a part of day-to-day life for businesses, and with the new technology it is creating a lot of opportunities. Kumar and Keshan (2009), Al-Mashari et al. (2006), and Molla and Bhalla (2006) in their papers showed the pathway for implementing the technology in organizations and its coming challenges.

10.3 Methodology and Analysis

The data that are used in this research were based on secondary data. Secondary data can be defined as data that are not collected by the research person, but it was collected by some other person or group. It is generally used when someone has already studied the subject you are analyzing earlier might be on same topic or any other different topic. In this research, data were collected from different sources, such as websites and other documents published by the company, to get a clear understanding of the topic.

The reason for using this method of data collection for this research was that the software on which we are doing our research is a professional software that is costly and used by big companies in their operations, so getting their knowledge of the software is not an easy task. So to get the information, global review sites were checked in which great users of the software give their important and detailed reviews for brainstorming questions.

For the analysis, the reviews were collected from different sources and were studied for determining the deployment of the software to different business units varying in size, type, and usefulness, which gave a wide scope for the study for the understanding of the software. Different factors were taken into consideration in the process of the analysis, such as the type of material and type of transportations used by Tata in their manufactured products, so that they could be reliable sources for the company. Data were ordered according to the requirements from all the data as that might be useful for the company because software was used by different manufacturers, and it cannot be applied to any company without a proper understanding of the data according to the business model.

10.4 Scope of the Study

The reason for the research was to know the efficiency of the Oracle transportation management software, when it is applied and used in the Tata Steel group. The main scope of the study of the software was to know what the pros and cons could be for the company while implementing it in a company.

The objective was to determine which critical criteria needed to be comprehended and which considerations must be taken into account when selecting software for the organisation, such as compatibility with the business's end outcomes. There were numerous elements to consider since the Indian business model is extremely different from that of other nations, and there are various geographical and human aspects to consider in India before bringing and implementing a technology in a corporation. Tata has done enormous research steps.

It studies the scope of the business and the scope of the software; how they can be an asset to the company with its different features; how they can be used to obtain a full return from the investment that was made on the software. As the software is quite very complex and easy to use, the main difference is whether the configuration has been done according to the requirements of the business or not, and how these can be achieved.

10.5 Analysis

10.5.1 Set Up and Configurations

The first and basic step that is to be taken by the company before using the Oracle Transportation Software[2] is the setting up of a process, as this is very important for the company because many configurations are required according to the requirements of the company. Different businesses have

[2] https://www.oracle.com/in/applications/supply-chain-management/logistics Acessed at 9 Pm on 25/04/2021.

different business requirements, and each requirement needs to be filled to full utilization of this software, because it is costly and very versatile software, so it needs to be used in a perfect manner to fulfill the requirements of the company.

There is much software integration available for various functions from procurement to warehousing to transportation to delivery (Figure 10.1).

For a company such as Tata Steel, this software can be a relief for their overall workload and also minimize the effort and time as they deal with their various sectors, such as

- Agriculture Equipment
- Automotive Steel
- Construction
- Energy and Power
- Consumer Goods
- Heavy Material Management

All these sectors are different from each other and require different types of shipping routes and shipping methods, so before implementation of the software, their requirements need to be analyzed and what software configuration needs to be used has to be decided for the full utilization of the software. If the company is using some software with regard to the working, there can be an integration of the software with the OTM.

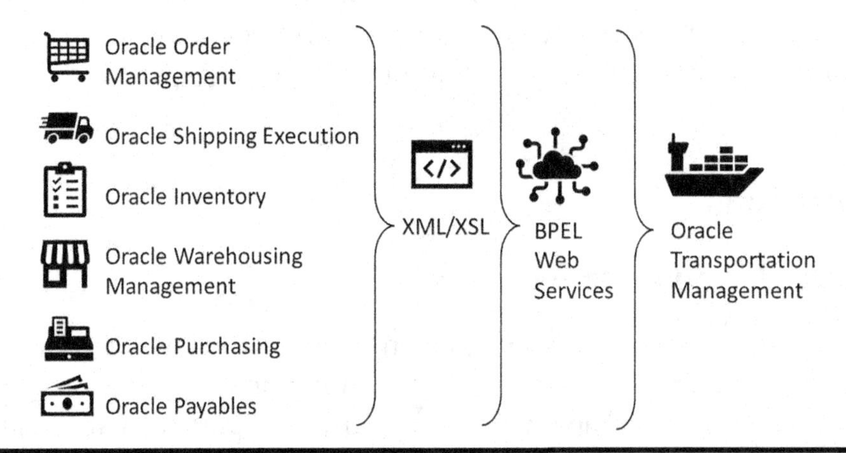

Figure 10.1 Oracle transportation cloud. (Authors' own compilation adapted from https://www.oracle.com/webfolder/s/quicktours/scm/gqt-scm-log-overview/index.html.)

10.5.2 Centralized Data Handling

As big firms that are working under different process lines under one parent need to maintain a lot of data and these data are needed in different departments and decision-making units, there should be an efficient flow of information among all so that it results in good on-time and future decisions. The data are not only required in the inter-office movement but also the inter-departmental movement of the company. Different departments need data to take different decisions with regard to the business, and it helps people to access them without any delays. If Tata wants to add something new to their supply chain network, then before starting the work, a proper analysis will be done, and all the data will be required by the planning as well as the financial departments to take up the cost and result analysis. With the centralized data stored in the cloud, they can retrieve data without any inter-department delays.

With the cloud data storage system, they can use the data to perform different simulations with different algorithms for their operations of the data; Oracle provides different forms of storage of data that can be used by Tata according to their requirement as per the requirement with a storage capacity of 32 TB to 1 PB.

For companies, such as Tata, two possible options are available. The first option is the "objective storage," where it will help the company to store data that is much required on a regular basis; it creates a dynamic movement of the data on a regular basis; it provides easy accessibility to the data in high time for better usage across platforms. It can support data, such as images, videos, and analyzed data sheets, to be accessed in real time. The second option is "archive storage," where it helps to keep data that are not much required frequently but could be required in the long run for any type of checkups or past data analysis for the company.

10.5.3 Multi-Tasking Functions

The supply chain is not a solitary operational entity; rather, it is one of the well-connected department-to-department and inter-department networks. Supply chain is not about the movement of the goods from one place to another; it is all about purchasing and delivering them to the right place. Tata has different working sectors and a lot of movement of goods and storage of goods; their daily part of the work when we see that not only manufacturing is important but its proper delivery and storage is also needed at

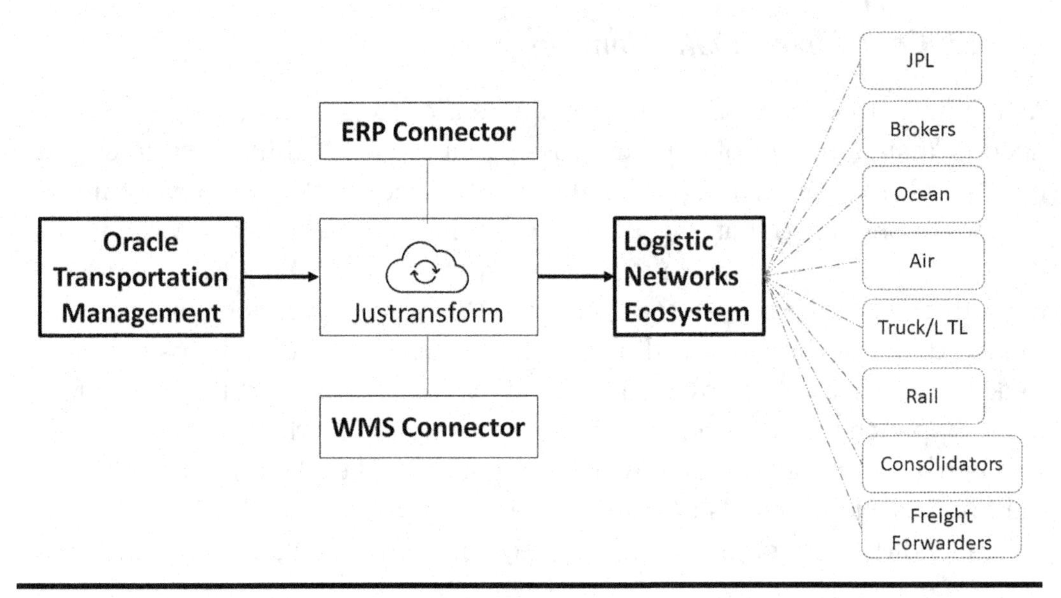

Figure 10.2 Oracle components and features. (Authors' own compilation adapted from https://www.oracle.com/in/scm/logistics/transportation-management/.)

the same time. To fight this problem, they can use different Oracle software functions available in the market (Figure 10.2).

Big companies can save a lot of time as well as money by utilizing their resources wisely for future planning. Companies can use OTM software with the integration of warehouses for the optimization of the warehouse areas and get organized data of all the products available with them. They can optimize their cost through the feature of rate management, which helps in getting the best and optimized cost for the company. They can manage their routes for the supply of their goods to the users or for the procurement of their goods from their other units; it helps in the end payments of the freight and billing of them for the records with these records and if any unexpected loss occurs insurance can be easily claimed.

These tasks can be operated by different people, but with this software, all of it can be done in a more integrated manner due to its cloud access system; every employee can be monitored with the current scenario and with that proper decisions can be made in real time.

10.5.4 Financial Aid

When we consider the logistics part of Tata Steel, they use logistics for their finished goods very heavily for their movement and delivery of goods; they use different means of transport from road to ship for their goods. With such usage of the logistics and supply chain departments, the company has a lot to face from damage to not able to use the full capacity of their means. This can be a great loss to the company if not fully managed, as it may end up in a situation where extra cost has to be met by their logistics department. It helps to design your vehicle from the booking zone to the goal region with its size, load, and complete limit available on the board for a better utilization of the cost as well as the space for transporting vehicles with no under-utilization of the resources.

It measures and controls financial performance with costing, payables, billable and dispute management, automate many manual procedures associated with freight bill payment and audit. It covers all aspects of the transportation rates, such as discounts and surcharges. It keeps track of the accessorial, match pay, auto pay, track claims, and report on cost. Oracle Transportation Management Cloud enables companies to manage all rates centrally, including support for truckload, intermodal, less-than-truckload, parcel, air, rail, and ocean. In the global sense, this software can help in the rate/route setup process by connecting the different service providers through different means and getting the best available option available to the company. It eliminates unnecessary charges and automates the time-consuming and error-prone processes of freight payment, customer billing, and managing freight claims.

10.5.5 Transportation Intelligence

In today's world, where AI is taking over companies and sectors, how can the emerging sectors of Indian economy stay away from it too? So Oracle Transportation Management can also be said as a step toward it too.

In simple words, it creates a transportation dashboard for operational and strategic performance indicators by leveraging the operational data stored in the Oracle Transportation Management Cloud. The software uses operational data from the cloud and converts it into meaningful data, which can be used in decision-making process. With this, one can avoid future inefficiency of the transportation and can make the needed changes for any plans by the company. It gives parameters that are the main reasons for the unsatisfactory results to the company so that needed changes can be applied to stop their hindrance. The software constantly uses the past recorded data and keeps doing its analysis for predicting future problems. The software analysis of different flexible indicators that define transportation operations is

- Orders, Shipments, Lines, and Stops
- Invoices and Invoice Lines
- Carrier Performance/Scorecards
- Sustainability
- Tracking Events
- Carrier Commitments
- Claims
- Quotes
- Drivers
- Planning and System Activity.

The main objective of AI is to provide the right information to the right person at the right time. It can create reports, matrices, and dashboards to support different operational decisions, which can be accessed by phones too. It helps in much of the decision-making from short to long goals of the company and helps not only the managers but also data analysts and transportation analysts by providing insights for the right decisions with some factual support.

10.5.6 Warehousing Guide

Warehousing can be defined as one of the most important part of the manufacturing units, as Tata being one of the major manufacturer of different industries for India needs to maintain its warehousing more efficiently and using its 100% capacity to stop the underutilization of their capacity; for this purpose Oracle has its warehousing management feature, which provides a complete visibility of the inventory to keep the mismatch of the inventory away. It keeps an eye on the inventory and schedule and manages inbound shipments, cross docking, which helps in improving the efficiency of the docks. It also configures, creates, and performs custom value-added services, such as labeling, tagging, and kitting at any time.

It can help Tata maintain their warehouses more efficiently, as this setup helps the inventory to be managed more efficiently and maintain the level of stock. It can be more efficient as it manages tasks of the distribution centers through mobile devices, including receiving, put away, order picking, replenishment, and truck loading. It also can help in real-time analysis for checkups through its options and capability of real-time reporting of data, which can be seen and sent to others on a real-time basis.

10.6 Shipment Management

As Tata is one of the leading manufacturers in the steel industry, it has a great grip on the global trade of the steel sector. Its steel not only is used in the country itself but also exported to other countries, Some countries are neighboring countries to India, so the loads can be transferred by road itself, but for countries such as UAE they generally use water routes, thus shipment service can also be seen as a vital service to the company logistics. With the help of the Oracle Global Trade Management software cloud the company can manage global customs process in one software; it can connect brokers, custom filling partners, or directly to the custom agencies to lodge declarations and clear the company's shipment (Kautish, Singh, Polkowski, Mayura & Jeyanthi, 2021).

The software runs automated decision support for full possibilities of the transportation scenarios from simple one-stop shipment to multi-station shipments, which provide operational efficiency to the company decisions (Singh, Singh & Karki, 2021). The software gives a robust functionality to manage the exceptions, and during the execution of the shipments it keeps track of the actual versus the planned movement of the ship, and it can also keep the diverting movement of the ship routes if any need is there.

10.7 Logistics Modeling

Tata Steel products can be basically seen as the raw products to many industries, and the efficiency of their logistics movement is a very important part of their brand, but with global movement, it can't be safe to rely on one network or on assumptions that might change according to the situations (Hofmann, Neukart & Bäck, 2017).

With the help of the Oracle logistic network modeling cloud, the company can see the different scenarios that it wishes to see and analyze in context to the real operational environment. There are different types of analyses available and can be used in every project and get multiple scenarios from the system. It provides us with different variations of data due to the scenario and can be used while deciding on the right policy.

The software provides multiple analysis on the same dashboard to understand the different variations from different scenarios. The scenario analytical dashboard gives a full analytical report that can be analyzed and compared to get a better understanding of the scenario under some common matrices, such as cost, utilization. With this we can define our matrices according to our needs, as Tata Steel can add the weight and size of the steel rods and other parts to be more efficient in the load movement. There are two different modes used by the software to analyze data, i.e., "strategic scenario analysis" and "tactical scenario analysis" modes.

Strategic scenario is helpful in the long-run goals of the firm with the impact on the business and seeing the long-term logistics network while other one work in coming up with the current situation problem optimization.

10.8 Interface

The Oracle software is quite complex to work with, as it needs proper guidance and proper configuration for a better use of the software. To use the software in all cases, its configuration has to be done correctly according to the needs of the business, and there are a lot of processes in the configuration that need to be completed before its use. The route path was not very logical and arranged to get to the correct setting or information, which poses an issue with obtaining the information. The user interface for Oracle is considered to be very lousy and not up to the mark; it works in a too old-fashioned manner on their website.

10.9 Conclusion

Today's business is not just a trade for the company; it is a strategy that is very important to survive in the market today. Every decision made in a business turns into a long-term profit or a long-term loss that can be recovered is not sure. With a diverse business that Tata is having, it is necessary to make decisions more wisely and with a proper analysis of the situation (Fritzsche, Kittel, Blankenburg & Vajna, 2012). This problem is very diverse as every environment that a business face is different from each other, and one of the emerging factors that is coming out in business is Logistic and Supply chain department that is a very important part of every manufacturer. As in a previous case study done by the Mukti Suraj and Dr. Rawani A shows a path of ERP implementation of SAP in the company in their research; they talk about all the challenges and the paths to implement the total ERP programs in the company, but with the limited technology in 2007–2010 it was seen as a big challenge. With technology enhancements in the nearby years in SAP and Oracle ERP softwares, it has become a lot easier due to enhanced studies and mindful knowledge among people.

Tata with such a large turnover and such a diversification in the products needs to be well maintained with data. This research was conducted to determine what the outcome could be of the implementation of Oracle Transportation Management software in different situations. Many different reviews were analyzed from different people in different sectors. The various understandings of the Oracle software that came out were that it is quiet advanced and needs to be worked upon smartly, as its confirmation requires different aspects of the business from business to business to utilize it perfectly.

This software helps in different aspects of the business from warehousing to route selection for transportation; it works on an end-to-end process with different analytical abilities to work on different data analyses to result in perfect decision-making for the business; it provides financial aid as well as optimal solutions to the business with its online intelligence. It works on past data analysis to provide results in a perfect and more optimized way by analyzing the past decisions and what their outcomes were and tries to give more optimal decisions. This software can handle a lot of work for the company and improve the workings of the overall operations smoothly. Even the software is a fully packed solution to many problems, it still lacks some of the factors, such as user interface, as it is seen that the user interface of the software is quiet old-fashioned and can be more efficient for the person interface and the after service of the software is quiet not well maintained by the company.

References

Al-Mashari, M., Ghani, S., & Al-Rashid, W. (2006). A study of the critical success factors of ERP implementation in developing countries. *International Journal of Internet & Enlevrise Management*, 4(1), 1–11.

Bagale, G.S., Vandadi, V.R., Singh, D. et al. (2021). Small and medium-sized enterprises' contribution in digital technology. *Annals of Operations Research*. Doi: 10.1007/s10479-021-04235-5.

Cheung, Y., & Bal, J. (1998). Process analysis techniques and tools for business improvement. *Business Process Management Journal*, 4(4), 274–290.

Davenport, T.H., & Short, J. (1990). The new industrial engineering: Information technology and business process redesign. *Sloan Management Review*, 34(4), 11–27.

Fritzsche, M., Kittel, K., Blankenburg, A., & Vajna, S. (2012). Multidisciplinary design optimization of a recurve bow based on applications of the autogenetic design theory and distributed computing. *Enterprise Information Systems*, 6(3), 329–343.

Groznik, A., & Maslaric, M. (2010). Achieving competitive supply chain through business process re-engineering: A case from developing country. *African Journal of Business Management*, 4(2), 140–148.

Hofmann, M., Neukart, F., & Bäck, T. (2017). Artificial Intelligence and data science in the automotive industry. *Data Science Blogs*, 489, 484–489.

Kautish, S. (2008). Online banking: A paradigm shift. *E-Business, ICFAI Publication, Hyderabad*, 9(10), 54–59.

Kautish, S., Singh, D., Polkowski, Z., Mayura, A., & Jeyanthi, M. (2021). *Knowledge Management and Web 3.0: Next Generation Business Models*. De Gruyter, Berlin.

Kautish, S., & Thapliyal, M.P. (2013). Design of new architecture for model management systems using knowledge sharing concept. *International Journal of Computer Applications*, 62(11), 11–13.

Kumar, S., & Keshan, A. (2009). ERP implementation in Tata steel: Focus on benefits and ROI. *Journal of Information Technology Case and Application Research*, 11, 68–103.

Molla, A., & Bhalla, A. (2006). Business transformation through ERP: Case of an Asian company. *Journal of Information Technology Cases and Research*, 8(1), 34–54.

Rejeb, A., Sűle, E., & Keogh, J.G. (2018). Exploring new technologies in procurement. *Transport & Logistics: The International Journal*, 18(45), 76–86.

Singh, A., & Gite, P. (2015). Corporate governance disclosure practices: A comparative study of selected public and private life insurance companies in India. *Apeejay - Journal of Management Sciences and Technology*, 2(2).

Singh, D., Singh, A., & Karki, S. (2021). Knowledge management and Web 3.0: Introduction to future and challenges. In *Knowledge Management and Web 3.0*. De Gruyter, Cambridge University Press. Doi: 10.1515/9783110722789-001Agents.

Smock, D., Rudzki, R.A., & Rogers, S.C. (2007). *On-Demand Supply Management: World Class Strategies, Practices, and Technology*. J. Ross Publishing, 9–11.

Tipping, A., & Kauschke, P. (2016). Shifting patterns: The future of the logistics industry. *PWC Publications*, 1, 1–17.

Umbenhauer, B. (2017) *Digital Supply Networks Strengthen Procurement Strategy*. Deloitte Publications, 1–11.

Chapter 11

New Marketing Perspective in Post-Covid Era with the Application of Business Intelligence

Nishi Pathak
Raj Kumar Goel Institute of Technology

Vishal Srivastava
JAIN (Deemed-to-be) University

Ashish Kumar Singh and Yatika Rastogi
Raj Kumar Goel Institute of Technology

Contents

DOI: 10.4324/9781003184928-11

11.1 Introduction

Coronavirus disease (COVID-19) is an infectious disease caused by coronavirus 2 that causes severe acute respiratory syndrome (SARS-CoV-2). It was discovered for the first time in China. COVID-19 was later labelled a pandemic by the World Health Organization on March 11, 2020, causing widespread worry. As no specific medicine or vaccine is available, more than 7.03 million people have become infected and 3.15 million people have recovered across 188 countries since November 2019. Social distancing, wearing masks, working from home, online learning, cashless payments, online marketing, and businesses are the new normal. There is also a huge change in the buying patterns of customers.

Amid the coronavirus pandemic, several countries have turned to lockdown to combat the infection. This lockdown means people are confined to their homes, have no movement on roads, work is restricted to laptops, Schools and colleges using e-learning platforms, industries shut, and the economy is experiencing its worst-ever depression. According to the International Monetary Fund, global economy will shrink by approximately 3%.

An article published in *New York Times* says, while some are shutting down their businesses, a few still find the right opportunity for them to start a new business. "Downturns or challenging times are to be seen as the right time to start a business", said Rashmi Menon, entrepreneur in residence at the University of Michigan. She mentions that as there is less competition for resources, the ups and downs of business whatever happens will at least give a new customer base.

The implications of this outbreak are far more resilient than we think. Still, the biggest challenge companies are facing is resuming operations.

With the fast-moving and unexpected variables, we must be prepared with some survival strategies. Also, keep a check on the customers' demand, variations in the current product, if any, according to the social sentiment trend (forbes.com), price, place of selling the product as well as promotion strategies of doing business. This outbreak led to impacting consumer mobility in choosing as well as in consumption of the products. From zero contact delivery to virtual product display and pay per click, advertising is the smart move to gain a competitive advantage. So, the marketers must plan for out-of-the-box marketing moves on screen as well as off-screen to capture the market in this extraordinary situation.

Thrive, a digital marketing company's SEO Manager says that "it is the great time to strengthen your online campaign." Search Engine Optimizations (SEO) and Google SERPs can help us to increase online traffic on our website. According to the article published in Forbes, March 19, 2020, a marketer should use Google query volume and quick response mechanisms to rebuild itself in the changing economic scenario. To flatten the infection curve and to make the economy's curve into a V shape, industries must mobilize this lockdown to strategize the future demands. New opportunities in the form of fewer travelling expenses, increased time and cost efficiency, and better connectivity paved the way to earning more profit on the businesses. Still, we must fight in a long way, as prolonged dependence on digital technology, loss of wage work to the unorganized sector, disturbed family life, and work–life balance are the dark sides of lockdown.

11.2 Research Strategy

The study is conducted keeping in mind the current pandemic situation with the following objectives:

1. To study the change in social activities and business operations forced by COVID-19
2. To study the technological advancements to serve changing market needs
3. To identify the area of changes in marketing operations

11.2.1 Methodology

The research used in the study is exploratory in nature. This study aims to understand the problem impact of the spread of COVID-19 over business practices in a better way. Three main aspects have been taken into consideration to study the above-mentioned objectives. They are

1. changes in the market structure
2. changes in business operations
3. technological advances in business intelligence.

The research work establishes a linkage between these three aspects to fulfil business objectives. The study referred to 50 research articles on the challenges imposed by COVID-19 over the restructuring of society and on redefining as well as redesigning corporate operations, from indexed reputed journals. To get better insight and establish interlinking among the above-mentioned domains, expert opinions were also considered.

11.2.2 Scope of This Research

The current research suggests the marketers, business owners to utilize this time in reinventing and reenergizing their operation to maximize productivity. This research broadens our perception of the costs and benefits of COVID-19. It is better to be acquainted than endangered. It is better to be proactive to secure our business (Kautish, 2008, Kautish and Thapliyal, 2013). Such pandemics could occur in the future too, but we must be prepared with some sort of survival strategy to fit with. This research explains the shift in the market opportunities, working styles, consumption patterns, and technological adaptability of customers due to this outbreak. Sooner or later, everything will recover, it is such that we must keep patience and open our imagination to accept the transformation. The era is of the survival of the fittest. The study will be useful for business people to rethink and redefine their businesses and help researchers for further research work.

11.2.3 Research Framework

The article identified the traits of the changing market along with the changes that are taking place in business operations. Considering the advancement happening in ICT, the article tries to give a workable marketing model with which organizations can attain their goals. The article maps the advanced ICT tools with the organizational and market needs to develop such a model. The research framework applied here is given below (Figure 11.1).

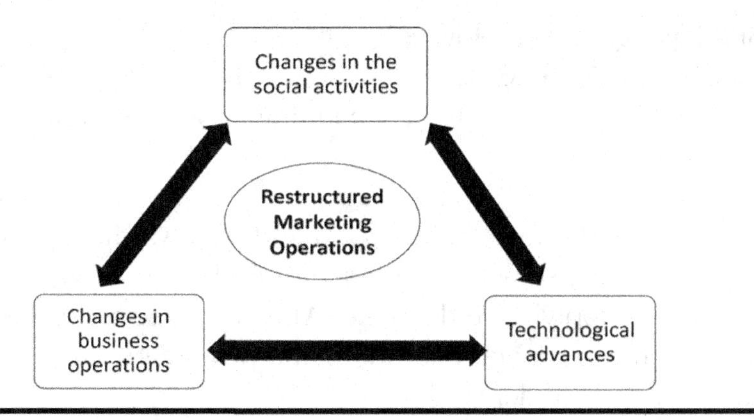

Figure 11.1 Research framework.

11.2.4 Discussion

The coronavirus spread has left almost all economies shattered. India is one of them, being on the fifth place (as of 9/6/2020) for the disruptions caused by the virus spread. The pandemic condition has affected the economy, society, businesses, and individuals. We have been seeing so many transformations: there is social distancing, wearing of a mask is compulsory, work from home culture is creeping in and to date, many of the workers are still in queue to return to their hometowns. We have not seen our near and dear ones almost for 2 months, thanks to the growing digital avenues at least we are virtually in touch with them.

The business sector is almost down facing the current pandemic situation. Many of them have been temporarily shut, leaving the workers unemployed. There are pay cuts, many well-known companies have started retrenching their staff due to low business.

- "ZOMATO cuts 13% jobs, rest face 50% pay"
  TOI-Pg 15, 16/05/2020
- "Swiggy lays off 1,100 employees"
  TOI-Pg 13, 19/05/2020
- "India bulls housing finance lays off several employees"
  TOI-Pg 15, 21/05/2020
- "OLA lets go 1/3rd of the work force"
  TOI-Pg 15, 21/05/2020
- "IBM's global lays off about 3,50,000 global strengths"
  TOI-Pg 17, 23/05/2020

- "UBER lays off 600 employees"
  TOI-Pg 11, 27/05/2020.
- "Book My Show lays off and furloughing approx. 19% staff" -TOI-Pg 19, 29/05/2020

It is all a miracle of the forces of demand and supply. There is no or very little demand from consumers as people are frightened of virus transmission. So, what will the supply force do alone? Almost all sectors, namely agricultural, aviation, entertainment, education, and hospitality, are seeing a downfall because of a lack of demand.

But are we really at a loss? Can't we reform our business practices to cope with the situation and make profits? In a recent address, the honourable Prime Minister Modi talked about the concept of 'Atmanirbhar Bharat'. This worst situation can also be an opportunity if efforts are put in the right direction. India, being a developing country, has the potential to produce more and more on its own land instead of importing from other countries. Even the established business will have to bring in a lot many changes for their survival. Today's customers are more available on digital platforms instead of physical platforms. They have to be pitched right there. The options, such as Say Namaste, Zoom, WebEx, Google Meet, Skype, etc., have made possible the virtual meetings with the customers without making any physical presence. Training for employees is nowadays on virtual platforms. All these things help with social distancing, which is the most important safety measure to avoid virus transmission.

Working from home culture wherever possible will help in reducing the operating cost of the business. It will give the employees a stress-free environment. More digital payment avenues usage will reduce the use of cash, will try making India a cashless economy and at the same time prevent the spread of the virus through currency circulation.

The environment is relieved and nature is healing itself during the lockdowns. People can spend more time with their families, which would have been impossible had it been a normal situation.

To summarize, this pandemic has brought in a lot of changes in the society and the economy. But we have already entered into this pandemic without an exit option to get out; it is now the time to revive our business practices according to the situation. It is now high time to benefit from the opportunities underlying and initiate our business start-ups to make the concepts of Make in India and Atmanirbhar Bharat possible. Digitalization is going to be the backbone for the success of all businesses shortly. India is a rich country with abundant skills and resources. Made in India will be the new enlightenment from now on, whether for a customer or a producer.

11.3 Change in Market Structure

COVID-19 has left many of the economies devastated. There has begun a transformation from being social to social distancing. The spread of the virus has brought in a lot of many changes in the society and the economy. The situation is somewhat like that which people faced after World War II. It is a health calamity declared by WHO. This virus is spreading all around the world, affecting health, employment, GDP, and social well-being without the consideration of class, colour, creed, or status. There is loss of life, loss of business, and loss of employment causing a setback to the emotions and psychology of the society. Spiritual sentiments are left aghast. Mobilization of the economies has come to a standstill. The economies are fearing inflation and the side effects of unemployment. The virus spread has restricted the mass gatherings.

The positive impact on society has been the healing of the environment. Nature is restoring, air is cleaner, and wildlife is at a bit of relief.

The agricultural sector is one of the prime sectors that has seen a lot of disturbances as the demand for agricultural commodities by hotels and restaurants has gone down by 20%. Sectors, such as manufacturing, education, finance, real estate, hospitality, sports, aviation, and entertainment, have all collapsed creating psychological pressure on the people of the society (Nicola Maria, 2020).

However, the major change in market structure is social distancing (Greenstone & Nigam, 2020). Considering many clinical and non-clinical studies, social distancing is uniformly accepted as nu normal. The research work of Allcott et al. (2020) highlighted the significance of social distancing in controlling the spread of the COVID-19 infections significantly. Wilder-Smith and Freedman (2020) stress Isolation, quarantine, social distancing, and community containment to fight the spread of COVID-19, such as epidemic and pandemic. Lewnard and Lo (2020) claimed in his research work- "the only strategy against COVID-19 is to reduce mixing of susceptible and infectious people".

This advocated social distancing led to a short fall in labour supply and labour (Barrot, Grassi, & Sauvagnat, 2020). The fear of the pandemic also worsens the market confidence (Depoux et al., 2020). All the above-mentioned factors led to gaps in food supply and distribution and thus imposed psychological pressure (Suresh, Barkatullah, & Boardman, 2020) on the market.

11.4 Change in Business Operations

The COVID-19 has greatly affected the labour market, even in big and developed economies, such as the United States. There has been a decrease in employment in upper and lower lines of labour by approximately 9% and 35%, respectively. The employment rate has gone down by almost 22% consisting of a decline of three times the new peak to trough changes that happened due to the Pandemic. Business suspensions have added to increased costs. The effect has created an alarming situation (Cajner et al., 2020).

COVID-19 has downsized all the economies (Bagale et al., 2021). There has been a macroeconomic effect, creating a deadly disaster in the history of the big economies. COVID-19, being a multidimensional shocker for the economy, has affected the forces of demand and supply and production altogether. But to focus more on long-term factors to be considered are economic and social well-being, the health of an individual, and the nation.

Ludvigson in his research work has considered two ingredients: the costly disaster and the deadly disaster. The effects of these factors have been studied on economic activity and macroeconomic uncertainty.

He has summarized that the COVID-19 has cost more damage in the form of a disaster that has affected the economies in comparison to the past decades.

The above study concludes that this disaster of COVID-19 has led to serious disturbances in industrial scenarios, as the pandemic has a different destroying nature from other disasters in the past 40 years (Ludvigson, Ma, & Ng, 2020).

The COVID-19 deadly pandemic has not left any of the economies unaffected. Nor has there been any sector left untouched by the cruel effect of the virus spread. This has led to several safety steps taken by the government, whether it is a lockdown, social distancing, or shutting down of small businesses temporarily. One of the researches works that were conducted confining to Malaysia focused on the effect of the safety measures implied to reduce the risk of Covid on SMEs' and their survival efforts.

Ahmad Raflis Che Omar in this study focusing on Malaysian SME's concluded that accumulated finances are a boon at this moment the tool to be financially dependent on. The other assets and skills available are being used to tap the new opportunities in the market for survival and growth. Several suggestions are provided by the respondent SMEs to speed up the survival strategies (Omar, Ishak, & Jusoh, 2020).

The rise and fall of any business in the 21st century is not only dependent on good and poor economic policies. The combination of man-made, human, and technical resources determines the sustainability of any business. For leading in business one of the most important tools is economic intelligence. It is the art of detecting uncontrollable factors by collecting, defining, and distributing information to the people (Anis, 2019).

The role of the customer in expanding business is not only restricted to purchasing the product. Customers' loyalty and satisfaction are the two driving forces for improving the financial performance of the company. Customers influence organizations' vision, mission, marketing plan, budget, expected future sales and prospects, organizational strategic management, marketing strategies in terms of pricing, value chain, and promotion strategies. As customer satisfaction is quite a challenging task, sometimes it leads to dissatisfaction (Mohammad et al., 2016). Customer psychological elements will always be kept in priority by the organization. So, customer satisfaction is important for the survival and sustainability of any organization. (NDU Chukwuogor Chiaku).

In the era of the 21st century, FDI plays a major role in the development of the nation. According to the UN (1997), FDI accounted for 25% of international capital flows. To fulfil the objective of wealth maximization, business engages them in FDI. It is creating more value for its shareholders. For attracting more FDI free flow of goods and services and regional economic integration is a necessity (Hamzah & Shamsudin, 2020).

The detailed literature review on the subject identified the following major changes going in business operations- focus on work from home (Beauregard, 2011; Hochschild, 2001). The labour scarcity induced cost inflation & corporates are suffering from stress on cost controlling (Smeets, Waldman, & Warzynski, 2019). However, the emergence of digital technology fuelled the business process automation (Rizk et al., 2020), this positively influencing the productivity & production quality (Gao, van Zelst, Lu, & van der Aalst, 2019). Augmented Reality (AR)- and Virtual Reality (VR)-assisted simulation or virtual training (Thies, Strohmeyer, Ebert, Stamminger, & Bauer, 2019) are widely used now. Even in this new normal, conducting virtual meetings is very common (Rubinger et al., 2020). But still, corporates need to consider work–life balance before setting the policies for work from home as in revised scenario the work from home extending working hours (Edge, Coffey, Cook, & Weinberg, 2020).

11.5 Technological Advances: Business Intelligence

In the past few years, it has been the growth years for digital avenues. We have been into the world of ICT. There has been constant reformation and revolution in communication and computing. The internet is at our fingertips now. We have become more dependent on digital platforms than physical ones. The business is now on the internet. Industrial society is becoming an information society converting the economy into a digital economy. The business can now reach out to more of the customers directly, easily, and conveniently. The business intelligence (BI) leaded advantages are increasing, opportunities are growing. Among the 4 Ps of marketing, Distribution and Promotion are now more dependent on the internet because the customers are on the internet. The communication medium is the internet for both urban and rural areas. Gone are the days when the business had to think twice to cater for the rural consumers. There is a creation of new business models. We have a virtual business, supported by BI, running simply on computers through the easy click of a mouse without even their physical existence. There is a creation of new markets through tapping of untapped markets especially in the rural areas with all praise to the Internet & BI (Berisha-Shaqiri, 2015). The proven advancement in BI and business analytics, which are making a huge change in business operations – e-retailing (Sarkar, 2016). It enabled online meeting platforms (Nayak, 2020) are now supported e-commerce (Agarwal, Ghosh, Li, & Ruan, 2019; Soutter, Ferguson, & Neubert, 2019; Schlosser, Lieberman, & Zheng, 2020; Raj, Jain, & Chauhan, 2020; Gupta, 2020) Digital Menu & VR (Farshid, Paschen, Eriksson, & Kietzmann, 2018; Jung, 2019; Fish, & Fujii, 2020), e-CRM (Ahmed, Amroush, & Maati, 2019; Al-Dmour, Algharabat, Khawaja, & Al-Dmour, 2019; Hamid, Mousavi, & Partovi, 2019; Rosalina & Malik, 2019) and e-SCM (Erceg & Damoska-Sekulowska, 2019; Akbar & Darius, 2019).

The dimensions identified from various research articles are summarized in the mentioned table below (Table 11.1).

Considering the literature survey, the study revised framework as given below (Figure 11.2).

Table 11.1 Literature Survey Summary

S. No.	*Research Aspect*	*Observation*	*Author*
1	Change in market structure	Social distancing	Greenstone and Nigam (2020), Allcott et al. (2020), Wilder-Smith and Freedman (2020), Lewnard and Lo (2020), Barrot, Grassi and Sauvagnat (2020)
		Social hygiene and isolation	Depoux et al. (2020), Williams, Armitage, Tampe, and Dienes (2020), Berg-Weger and Morley (2020), Cudjoe et al. (2020)
		Psychological pressure	Suresh, Barkatullah and Boardman (2020), Downing, Ahmed, Vidal-Alaball, and Lopez Seguí (2020), Niles et al. (2020), Pérez-Escamilla, Cunningham, and Moran (2020)
		Virtual presence with the family	George (2020), Hülür and Macdonald (2020), Kravchenko (2020)
2	Change in business operations	Focus on work from home	Beauregard (2011), Hochschild (2001), Dikkers et al. (2007), Nippert-Eng (2008), Patton (2019), Jha (2019)
		Stress on cost controlling	Sharma (2019), Gurvich, Lariviere, and Moreno (2019), Smeets, Waldman, and Warzynski (2019), Isaksson, Harjunkoski, and Sand (2018)
		Business process automation	Rizk et al. (2020); Gao, van Zelst, Lu, and van der Aalst (2019), Madakam, Holmukhe, and Jaiswal, (2019)
		Using virtual trainings	Thies, Strohmeyer, Ebert, Stamminger, and Bauer (2019), Menke, Beckmann, and Weber (2019), Bayerl, Davey, Lohrmann, and Saunders (2019), Ziegler, Papageorgiou, Hirschi, Genovese, and Christ (2020)
		Conducting virtual meetings	Rubinger et al. (2020), Mori (2020), Liu, Castronovo, Messner, and Leicht (2020), Porpiglia et al. (2020), Jones and Abdelfattah (2020)
		Extending working hours	Edge, Coffey, Cook, and Weinberg (2020), Trudel et al. (2020), Ogg and Rašticová (2020), Lain, van der Horst, and Vickerstaff (2020).

(*Continued*)

Table 11.1 (*Continued*) Literature Survey Summary

S. No.	*Research Aspect*	*Observation*	*Author*
3	BI and BA advance-ments in Techno-logical	e-Retailing	Arora (2013); Krafft and Mantrala (2006); Sarkar (2016), Dixit (2016), Nayak (2020), Sheridan, Banzer, Pradzinski, and Wen (2020); Yoshioka et al. (2019); Bergmann, Braunes, and Flohr (2020).
		e-Commerce	Soutter, Ferguson, and Neubert (2019); Schlosser, Lieberman and Zheng (2020), Raj, Jain, and Chauhan (2020), Gupta (2020), Agarwal, Ghosh, Li, and Ruan (2019)
		Digital menu and VR	Fish and Fujii (2020), Farshid, Paschen, Eriksson, and Kietzmann (2018), Jung (2019)
		e-CRM	Ahmed, Amroush, and Maati (2019), Al-Dmour, Algharabat, Khawaja, and Al-Dmour (2019), Hamid, Mousavi, and Partovi (2019), Rosalina and Malik (2019)
		e-SCM	Erceg and Damoska-Sekulowska (2019), Akbar and Darius (2019), Chong, Bian, and Zhang (2016)

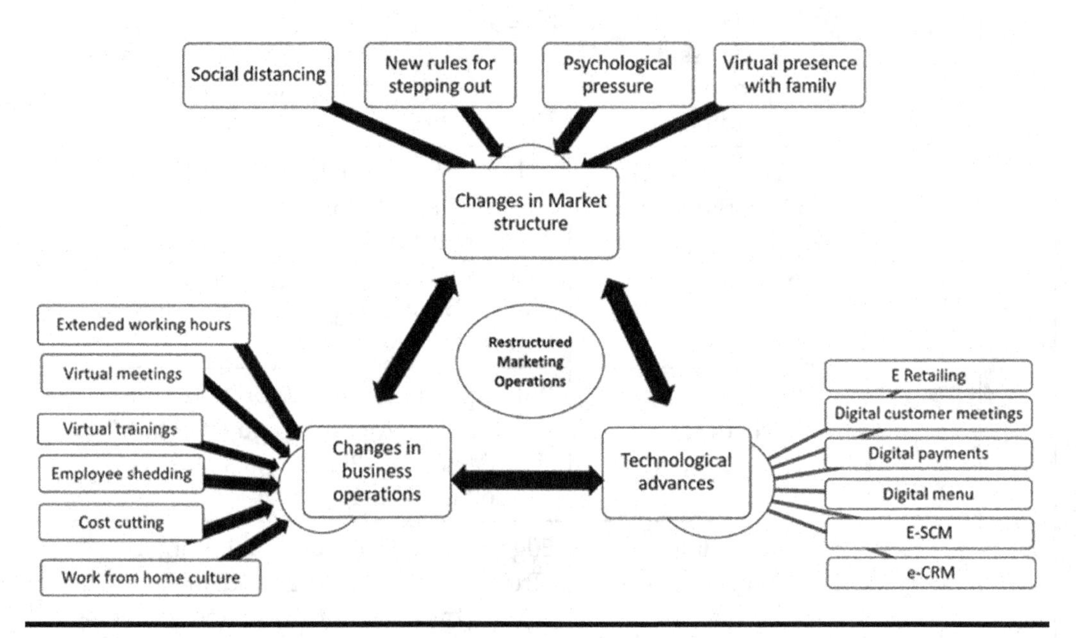

Figure 11.2 Revised research framework.

11.5.1 Analysis and Interpretation

The table is given below (Table 11.2), explains clearly that BI/ BA technological advancements are no doubt able to fulfil the need not only of changing business scenario but also of changing society.

Whether it is E-Retailing, Digital customer interaction, Digital Payments, Digital Menu, E CRM or SCM, all have been affected by the changes happening in the business and society due to the spread of coronavirus. Work from home has become an easy option as things are being operated digitally. Almost the jobs are 24×7 now, increasing the working hours just because of easy virtual availability. Customer meetings and sales virtually is another major change now. The products are being displayed virtually by salespeople now. When we see the interaction of BI with society can be seen. The use of digital options is helping in maintaining social distance, which is the need of the hour. It is creating more psychological pressure due to the increasing working hours. Of course, BI has paved a way for virtual presence with the family (Kautish, Singh, Polkowski, Mayura & Jeyanthi, 2021).

Moreover, to check the market need fulfilment, the study mapped society needs with technological competency of reframing business operations, it is found that the advanced business analytics/BI are competent enough to meet the requirements (Figure 11.3).

11.6 Research Implications: Changes in Marketing Operations

After analysing the application & usage of advanced BI/business analytics in achieving operations' goal of business and needs of society, the study identifies the execution of 4 Ps of marketing (Product, Price, Place & promotion) in changed scenario (Figure 11.4).

The opportunities identified during and post-COVID-19 situations concerning marketing functions are given below.

11.6.1 Product Decisions

It is found that innovative BI assists better than traditional ones in designing new products or services. The ICT helps a lot to easily connect not only with the target market to assess their needs but also with the experts and vendors to test the feasibility. The online meeting tools, digital display & simulations help a lot in such activities.

Table 11.2 Interaction of Business Intelligence/Business Analytics with Society and the Business World

	Business Change						*Social Change*			
Business Intelligence	*Work from Home Culture*	*Cost-Cutting*	*Employee Shedding*	*Virtual Training*	*Virtual Meeting*	*Increase in Working Hours*	*Social Distancing*	*Psycho-logical Pressure*	*New Rules to Step out*	*Virtual Presence with the Family*
E – retailing	x	x	x			x	X		x	x
Digital customer meetings	x	x		x	x	x	X	x		x
Digital payments	x	x		x	x	x	X		x	x
Digital menu		x				x	X	x	x	x
E CRM	x	x	x	x	x	x	X	x		x
E SCM	x	x	x	x	x	x	X	x	x	

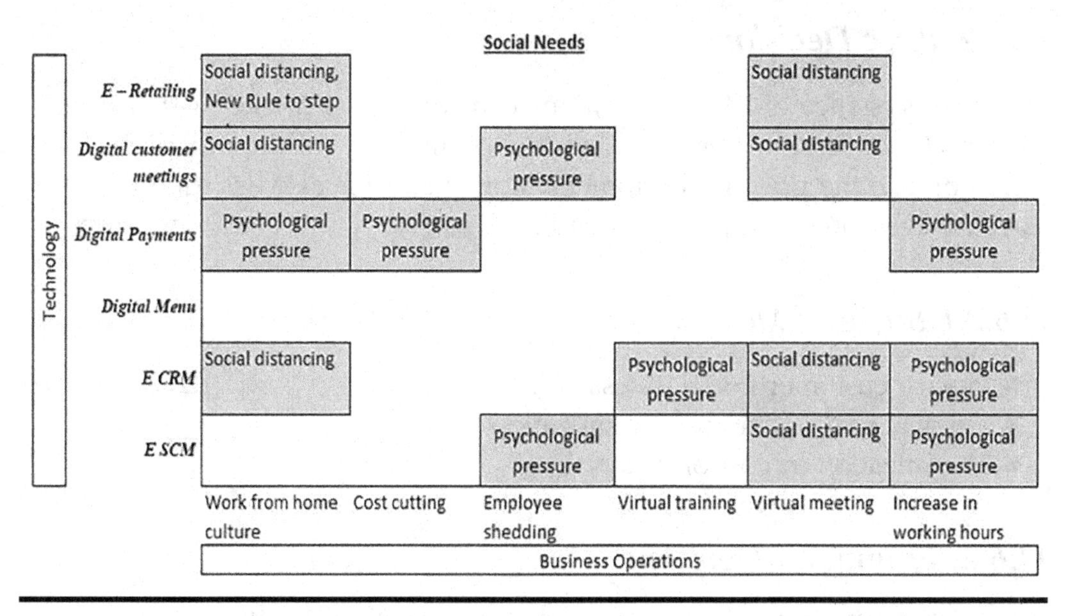

Figure 11.3 Mapping BI/ BA technological advancement and business operation in fulfilling marketing need.

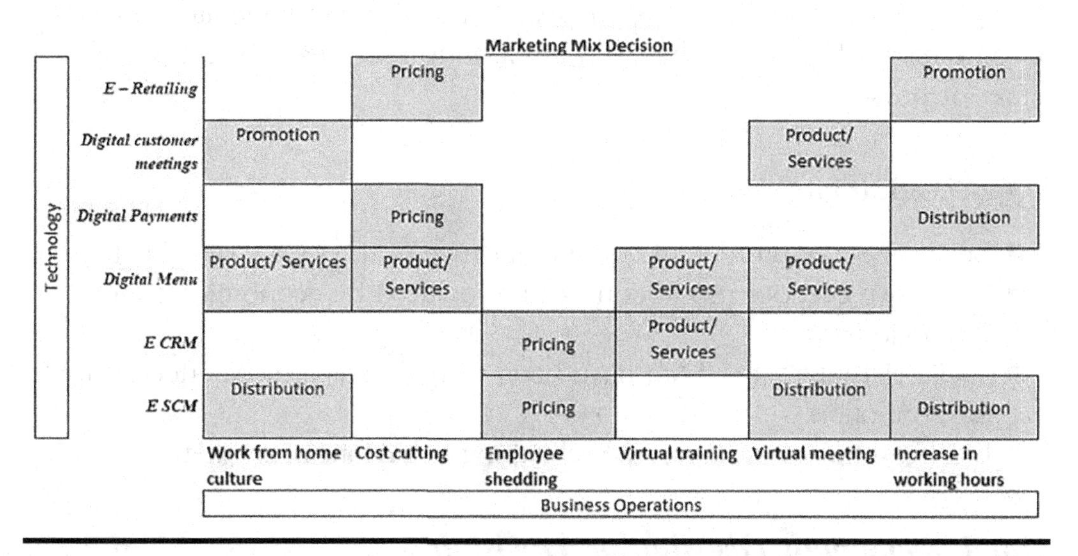

Figure 11.4 Strategic restructuring in marketing functions.

11.6.1.1 Strategic Move

- More customer involvement in product/ services designing
- Application of simulation and meeting integration platform in product development.

11.6.2 Price Decisions

With the assistance of BI, organizations can serve their market with a minimum cost, which ultimately increases the profitability for the organization and minimize the price for customers. Moreover, BI is also assisting in reaching everyone with pace and ease.

11.6.2.1 Strategic Move

- Precise customer needs assessment
- Physical movement cost minimization
- Minimization in cost of delay

11.6.3 Promotion Decision

During the COVID-19 situation, people are forced to spend more time either with their family members or on social networking sites. This time people are found more connected with their distant reference groups also through the internet. The study suggests that social network sites have become more effective media for promotion to reach each & every customer or prospect.

11.6.3.1 Strategic Move

- Below the line promotion is more effective in the COVID-19 situation
- Expensive outdoor promotion can be replaced by economics online promotion
- Social Network Sites (SNS) have been identified as a more effective tool for promotion
- BI offers the real-time trend & changing behaviour of market

11.6.4 Placement/Distribution Decision

To avoid personal contact, the vendor, supplier and intermediaries are also now more dependent on telephonic and online communication measures. The BI lead to minimizing the logistics assistance cost and digital monitoring of goods movement. This digitalization of logistics is speedier than the traditional one.

11.6.4.1 Strategic Move

- Tighter monitoring and more customized logistic operation
- Involvement of BI platform equipped with analytical tool offers more effective and efficient monitoring and distribution.

11.7 Conclusion

In this research paper, we have used three aspects: change in market structure, change in business opportunities, and technological advances in BI/BA to set a linkage between the three. Due to social distancing and restricted movement the society is dependent more on online revenues. The organizations are seeking to reduce costs and are induced to work with minimum employees as there is low demand. BI has come into the lime-light, with the latest business analytical tools coming in to stay connected virtually, maintaining the norm of social distancing. The producers, consumers, and all stakeholders are convinced and ready to connect virtually with the intimation tool giving advantage for distance, designing of products and services are becoming easier from all nook and corners of the nation.

There will now be less usage of office space contributing to reducing operational costs. Work from home culture will have less movement, reducing psychological pressure of the people and creating a better work environment.

The distribution will be equipped with E SCM involving lesser work-force requirements, distribution and logistics will be managed digitally. The inbound movement of goods will be managed through automation and outbound movement can be monitored through the internet. Promotion will be more on virtual platforms that will be used by the majority of the organizations, and that is the need of the hour. Connection through social networking sites will be a common thing. It is easier and faster to reach out to customers on digital platforms. By using BI, an organization will be able to serve the customers in a better and convenient way. Brand Awareness through digital avenues can be done more creatively. The promotion and information distribution can take place directly to customers at minimum cost.

The environment has also benefited and has been in a healing stage. There is a lack of pollution, we can see clear sky, chirping of birds is soothing; the greenery is giving calmness to eyes. So, as we have entered a state where we need to be distant socially, the business will have to reform themselves and connect on digital platforms for all activities for survival and profit at the same time.

11.8 Limitations

This study is based on 50 research articles, extensive literature survey can be done to get more precise outcomes. This is a highly turbulent phase therefore many more activities and approaches are being tested by different corporates and scholars to come out with few more additional effective and efficient strategies. Whatever the activity and strategy identified need to be tested to establish validity and reliability.

References

Agarwal, S., Ghosh, P., Li, J., & Ruan, T. (2019). *Digital Payments and Consumption: Evidence from the 2016 Demonetization in India.*

Ahmed, B. S., Amroush, F., & Maati, M. B. (2019). The intelligence of E-CRM applications and approaches on online shopping industry. In *Advanced Methodologies and Technologies in Digital Marketing and Entrepreneurship* (pp. 70–82). IGI Global.

Akbar, M., & Darius, A. (2019). E-supply chain management value concept for the palm oil industry. *Jurnal Sistem Informasi*, 15(2), 15–29.

Al-Dmour, H. H., Algharabat, R. S., Khawaja, R., & Al-Dmour, R. H. (2019). Investigating the impact of ECRM success factors on business performance. *Asia Pacific Journal of Marketing and Logistics*, Vol 2, 14–16

Allcott, H., Boxell, L., Conway, J., Gentzkow, M., Thaler, M., & Yang, D. Y. (2020). *Polarization and Public Health: Partisan Differences in Social Distancing During the Coronavirus Pandemic.* NBER Working Paper, (w26946).

Anis, A., & Susdarwono, E. T. (2020). 21st century skills of economic intelligence related to the resilience of the national economy. *Jurnal Pendidikan Ilmu Sosial*, 29(2), 127–139.

Arora, J. (2013). Prospect of e-retailing in India. *IOSR Journal of Computer Engineering*, 10(3), 11–15.

Bagale, G. S., Vandadi, V. R., Singh, D., Sharma, D. K., Garlapati, D. V. K., Bommisetti, R. K., … Sengan, S. (2021). Small and medium-sized enterprises' contribution in digital technology. *Annals of Operations Research*. Doi: 10.1007/s10479-021-04235-5.

Barrot, J. N., Grassi, B., & Sauvagnat, J. (2020). *Sectoral Effects of Social Distancing.* Available at SSRN.

Bayerl, P. S., Davey, S., Lohrmann, P., & Saunders, J. (2019). Evaluating serious game trainings. In *Serious Games for Enhancing Law Enforcement Agencies* (pp. 149–169). Cham: Springer.

Beauregard, T. A. (2011). Direct and indirect links between organizational work–home culture and employee well-being. *British Journal of Management*, 22(2), 218–237.

Bergmann, L., Braunes, J., & Flohr, D. (2020). U.S. Patent No. 10,542,238. Washington, DC: U.S. Patent and Trademark Office.

Berg-Weger, M., & Morley, J. E. (2020). Loneliness and social isolation in older adults during the Covid-19 pandemic: Implications for gerontological social work. *The Journal of Nutrition, Health & Aging*, 24(5), 456–458.

Berisha-Shaqiri, A. (2015). Impact of information technology and internet in businesses. *Information Technology*, Q2.

Cajner, T., Crane, L. D., Decker, R. A., Grigsby, J., Hamins-Puertolas, A., Hurst, E., … Yildirmaz, A. (2020). *The US Labor Market During the Beginning of the Pandemic Recession (No. w27159).* National Bureau of Economic Research.

Chong, W. K., Bian, D., & Zhang, N. (2016). E-marketing services and e-marketing performance: The roles of innovation, knowledge complexity and environmental turbulence in influencing the relationship. *Journal of Marketing Management*, 32(1–2), 149–178.

Cudjoe, T. K., Roth, D. L., Szanton, S. L., Wolff, J. L., Boyd, C. M., & Thorpe Jr, R. J. (2020). The epidemiology of social isolation: National health and aging trends study. *The Journals of Gerontology: Series B*, 75(1), 107–113.

Depoux, A., Martin, S., Karafillakis, E., Preet, R., Wilder-Smith, A., & Larson, H. (2020). The pandemic of social media panic travels faster than the COVID-19 outbreaks. *Journal of Travel Medicine*, 27(3), taaa031.

Dikkers, J. S., Geurts, S. A., Dulk, L. D., Peper, B., Taris, T. W., & Kompier, M. A. (2007). Dimensions of work–home culture and their relations with the use of work–home arrangements and work–home interaction. *Work & Stress*, 21(2), 155–172.

Dixit, S. (Ed.). (2016). *E-retailing Challenges and Opportunities in the Global Marketplace.* IGI Global.

Downing, J., Ahmed, W., Vidal-Alaball, J., & Lopez Seguí, F. (2020). Battling fake news and (in) security during COVID-19. *E-International Relations.*

Edge, C. E., Coffey, M., Cook, P. A., & Weinberg, A. (2020). Barriers & facilitators to extended working life: A focus on a predominately female ageing workforce. *Ageing & Society*, 41, 2867–2887.

Erceg, A., & Damoska-Sekulowska, J. (2019). E-logistics and e-SCM: How to increase competitiveness. *LogForum*, 15(1), 11–16

Farshid, M., Paschen, J., Eriksson, T., & Kietzmann, J. (2018). Go boldly! Explore augmented reality (AR), virtual reality (VR), and mixed reality (MR) for business. *Business Horizons*, 61(5), 657–663.

Fish, K. A., & Fujii, J. M. (2020). *U.S. Patent No. 10,663,755.* Washington, DC: U.S. Patent and Trademark Office.

Gao, J., van Zelst, S. J., Lu, X., & van der Aalst, W. M. (2019, October). Automated robotic process automation: A self-learning approach. In *OTM Confederated International Conferences "On the Move to Meaningful Internet Systems"* (pp. 95–112). Springer, Cham.

George, É. (Ed.). (2020). *Digitalization of Society and Socio-political, Issues 1: Digital, Communication and Culture*. John Wiley & Sons.

Greenstone, M., & Nigam, V. (2020). *Does Social Distancing Matter?* University of Chicago, Becker Friedman Institute for Economics Working Paper, (2020–26).

Gupta, D. (2020). Digital payments in India: A conceptual study. *Studies in Indian Place Names*, 40(3), 5906–5914.

Gurvich, I., Lariviere, M., & Moreno, A. (2019). Operations in the on-demand economy: Staffing services with self-scheduling capacity. In *Sharing Economy* (pp. 249–278). Springer, Cham.

Hamid, A. B. A., Mousavi, S. B., & Partovi, B. (2019). *Managing E-CRM Towards Customer Satisfaction and Quality Relationship*. Singapore: Partridge Publishing Singapore.

Hamzah, A. A., & Shamsudin, M. F. (2020). Why customer satisfaction is important to business? *Journal of Undergraduate Social Science and Technology*, 1(1).

Hochschild, A. R. (2001). *The Time Bind: When Work Becomes Home, and Home Becomes Work* (Vol. 2). Macmillan.

Hülür, G., & Macdonald, B. (2020). Rethinking social relationships in old age: Digitalization and the social lives of older adults. *American Psychologist*, 75(4), 554.

Isaksson, A. J., Harjunkoski, I., & Sand, G. (2018). The impact of digitalization on the future of control and operations. *Computers & Chemical Engineering*, 114, 122–129.

Jha, R. (2019). Understanding the culture of telecommuting and employee performance. *NOLEGEIN-Journal of Information Technology & Management*, 17–22.

Jones, R. E., & Abdelfattah, K. R. (2020). Virtual interviews in the era of COVID-19: A primer for applicants. *Journal of Surgical Education*, 77(4), 733–734.

Jung, T. (2019). *Augmented Reality and Virtual Reality: The Power of AR and VR for Business*. Springer Nature, Switzerland AG.

Kautish, S. (2008). Online banking: A paradigm shift. *E-Business, ICFAI Publication, Hyderabad*, 9(10), 54–59.

Kautish, S., Singh, D., Polkowski, Z., Mayura, A. & Jeyanthi, M. (2021). *Knowledge Management and Web 3.0: Next Generation Business Models*. Berlin: De Gruyter.

Kautish, S., & Thapliyal, M.P. (2013). Design of new architecture for model management systems using knowledge sharing concept. *International Journal of Computer Applications*, 62(11), 9–11

Krafft, M., & Mantrala, M. K. (2006). *Retailing in the 21st Century*. Heidelberg: Springer Berlin.

Kravchenko, S. (2020). *The Digitalization of Socium: Its Side Effects on the Socialization of the Young People*. Available at SSRN 3518317.

Lain, D., van der Horst, M., & Vickerstaff, S. (2020). Extended working lives: Feasible and desirable for all? In *Current and Emerging Trends in Aging and Work* (pp. 101–119). Cham: Springer.

Lewnard, J. A., & Lo, N. C. (2020). Scientific and ethical basis for social-distancing interventions against COVID-19. *The Lancet. Infectious Diseases*, 20, 631–633.

Liu, Y., Castronovo, F., Messner, J., & Leicht, R. (2020). Evaluating the impact of virtual reality on design review meetings. *Journal of Computing in Civil Engineering*, 34(1), 04019045.

Ludvigson, S. C., Ma, S., & Ng, S. (2020). *Covid19 and the Macroeconomic Effects of Costly Disasters (No. w26987)*. National Bureau of Economic Research.

Madakam, S., Holmukhe, R. M., & Jaiswal, D. K. (2019). The future digital work force: Robotic process automation (RPA). *JISTEM-Journal of Information Systems and Technology Management*, 16.

Menke, K., Beckmann, J., & Weber, P. (2019). Universal design for learning in augmented and virtual reality trainings. In *Universal Access Through Inclusive Instructional Design: International Perspectives on UDL*, 294.

Mori, A. S. (2020). Next-generation meetings must be diverse and inclusive. *Nature Climate Change*, 10, 1–1.

Nayak, P. (2020). *U.S. Patent No. 10,542,126*. Washington, DC: U.S. Patent and Trademark Office.

Niles, M. T., Bertmann, F., Belarmino, E. H., Wentworth, T., Biehl, E., & Neff, R. A. (2020). *The Early Food Insecurity Impacts of COVID-19*. medRxiv.

Nippert-Eng, C. E. (2008). *Home and Work: Negotiating Boundaries through Everyday Life*. Chicago, IL: University of Chicago Press.

Ogg, J., & Rašticová, M. (2020). Introduction: Key issues and policies for extending working life. In *Extended Working Life Policies* (pp. 3–27). Cham: Springer.

Omar, A. R. C., Ishak, S., & Jusoh, M. A. (2020). The impact of Covid-19 movement control order on SMEs' businesses and survival strategies. *Geografia-Malaysian Journal of Society and Space*, 16(2).

Patton, E. (2019). Where does work belong? Home-based work and communication technology within the American middle-class postwar home. *Technology and Culture*, 60(2), 523–552.

Pérez-Escamilla, R., Cunningham, K., & Moran, V. H. (2020). COVID-19, food and nutrition insecurity and the wellbeing of children, pregnant and lactating women: A complex syndemic. *Maternal & Child Nutrition*, e13036.

Porpiglia, F., Amparore, D., Autorino, R., Checcucci, E., Cooperberg, M. R., Ficarra, V., & Novara, G. (2020). Traditional and virtual congress meetings during the COVID-19 pandemic and the post-COVID-19 era: Is it time to change the paradigm? *European Urology*, 78, 301.

Raj, A., Jain, N., & Chauhan, S. S. (2020). Digital payments and its security. *Cybernomics*, 2(2), 13–20.

Rizk, Y., Bhandwalder, A., Boag, S., Chakraborti, T., Isahagian, V., Khazaeni, Y., … Unuvar, M. (2020). A unified conversational assistant framework for business process automation. *arXiv preprint arXiv:2001.03543*.

Rosalina, V., & Malik, A. (2019). Electronic Customer Relationship Management (E-CRM) modeling on micro, small & Medium Enterprises (MSMEs) Banten.

Rubinger, L., Gazendam, A., Ekhtiari, S., Nucci, N., Payne, A., Johal, H., … Bhandari, M. (2020). Maximizing virtual meetings and conferences: A review of best practices. *International Orthopaedics*, 44, 1–6.

Sarkar, R. (2016). E-Retailing: Boon or bane? *International Journal in Management & Social Science*, 4(2), 206–213.
Schlosser, R. A., Lieberman, A., & Zheng, M. Y. (2020). *U.S. Patent No. 10,540,646*. Washington, DC: U.S. Patent and Trademark Office.
Sharma, H. (2019). Cost controlling methodology: Design to cost (No. 2019-26-0077). SAE Technical Paper.
Sheridan, K. M., Banzer, D., Pradzinski, A., & Wen, X. (2020). Early math professional development: Meeting the challenge through online learning. *Early Childhood Education Journal*, 48(2), 223–231.
Singh, D., Singh, A., & Karki, S. (2021). Knowledge management and Web 3.0: Introduction to future and challenges. In *Knowledge Management and Web 3.0*. De Gruyter, Cambridge University Press. Doi: 10.1515/9783110722789-001Agents.
Smeets, V., Waldman, M., & Warzynski, F. (2019). Performance, career dynamics, and span of control. *Journal of Labor Economics*, 37(4), 1183–1213.
Soutter, L., Ferguson, K., & Neubert, M. (2019). Digital payments: Impact factors and mass adoption in Sub-Saharan Africa. *Technology Innovation Management Review*, 9(7), 11–13.
Suresh, A., Barkatullah, A. F., & Boardman, M. B. (2020). Addressing food insecurity during COVID-19: A role for rural federally qualified health centers.
Thies, L., Strohmeyer, C., Ebert, J., Stamminger, M., & Bauer, F. (2019, October). Compiling VR/AR trainings from business process models. In *2019 IEEE International Symposium on Mixed and Augmented Reality Adjunct (ISMAR-Adjunct)* (pp. 181–186). IEEE.
Trudel, X., Brisson, C., Gilbert-Ouimet, M., Vézina, M., Talbot, D., & Milot, A. (2020). Long working hours and the prevalence of masked and sustained hypertension. *Hypertension*, 75, 532–538.
Wilder-Smith, A., & Freedman, D. O. (2020). Isolation, quarantine, social distancing and community containment: Pivotal role for old-style public health measures in the novel coronavirus (2019-nCoV) outbreak. *Journal of Travel Medicine*, 27(2), taaa020.
Williams, S. N., Armitage, C. J., Tampe, T., & Dienes, K. (2020). Public perceptions and experiences of social distancing and social isolation during the COVID-19 pandemic: A UK-based focus group study. *MedRxiv*.
Yoshioka, T., Abramovski, I., Aksoylar, C., Chen, Z., David, M., Dimitriadis, D., & Hurvitz, A. (2019). Advances in online audio-visual meeting transcription. *arXiv preprint arXiv*:1912.04979.
Ziegler, C., Papageorgiou, A., Hirschi, M., Genovese, R., & Christ, O. (2020, April). Training in immersive virtual reality: A short review of presumptions and the contextual interference effect. In *International Conference on Human Interaction and Emerging Technologies* (pp. 328–333). Cham: Springer.

Chapter 12

The Application of Text Mining in Detecting Financial Fraud: A Literature Review

Pratibha Maurya
University of Delhi

Anurag Singh
University of Petroleum and Energy Studies (UPES)

Mohd. Salim
Al-Tareeka Management Studies (ATMS)

Contents

12.1 Introduction

Text is a common means of data exchange in the modern world. Text mining encompasses a variety of subfields, including natural language processing (NLP), information retrieval, web mining, computational linguistics, data extraction, and data mining. Automated structured data extraction

DOI: 10.4324/9781003184928-12

from unstructured and semi-structured materials was accomplished through the use of text mining (Kautish, 2008, Kautish and Thapliyal, 2013). Commercially, it is rather valuable. It is a novel technique for analysing massive sets of formless documents with the goal of extracting knowledge or non-trivial patterns. Document files come in a variety of forms, including text files, flat files, and PDF files. These files were assembled from a number of sources, including message boards, newsgroups, emails, online chat, text messages, and websites (Bagale et al., 2021). Humans are capable of rapidly resolving problems and of identifying and applying linguistic patterns to text (Singh & Gite, 2015). On the other hand, computers are incapable of handling difficulties, such as spelling, context, slang, and variation. Nonetheless, our language abilities and computing capabilities enable us to analyse text quickly or in enormous quantities in order to grasp unstructured data. A computer can analyse unstructured data using the text-mining technique. Fraud detection is a priority for financial sector organisations (Figure 12.1).

Financial fraud is a serious commercial concern on a global scale. Financial fraud involves a broad range of diverse sorts of deception, including

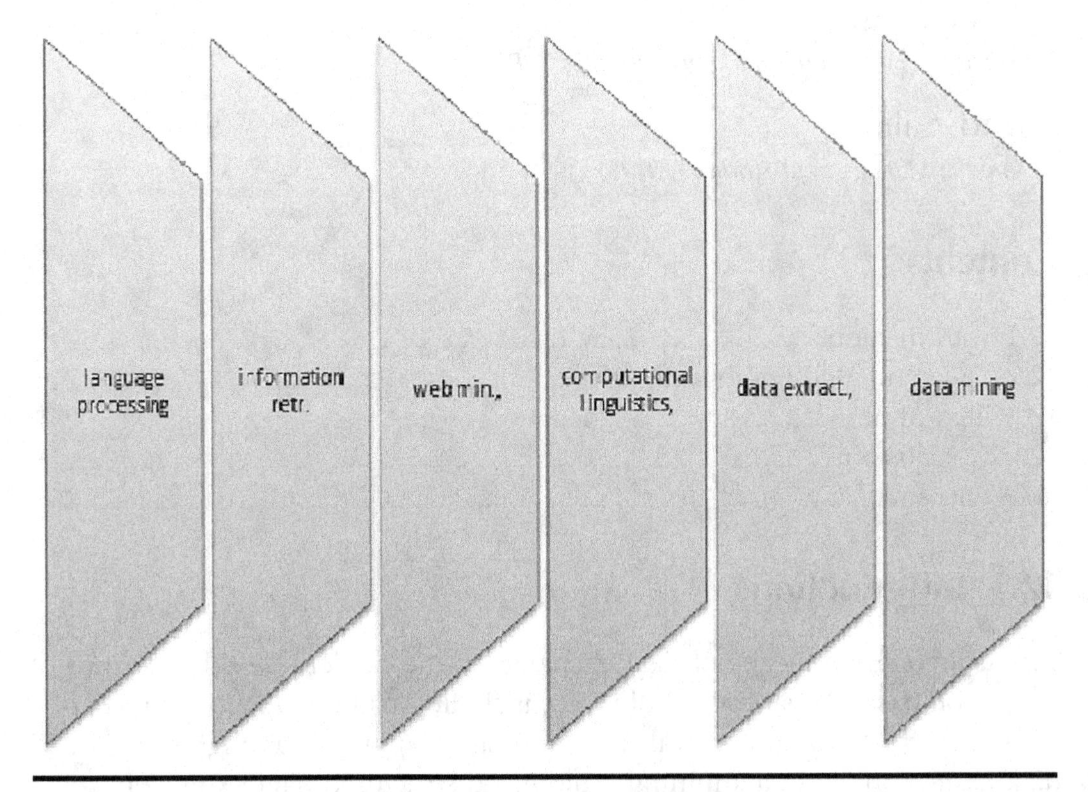

Figure 12.1 Contributors to the discipline of text mining.

healthcare fraud, mortgage fraud, corporate fraud, securities and commodities fraud, and fraud against financial institutions. When fraud protection fails, fraud detection is activated (Kautish, Singh, Polkowski, Mayura & Jeyanthi, 2021). In some cases, fraudsters are oblivious, rendering fraud prevention useless. As a result, ongoing fraud detection is required.

Financial statement fraud happens when an organisation's management gives fraudulent financial information (Singh, Singh & Singh, 2022). Financial statement audits ensure that the financial statements of a business are free of substantial misrepresentation and fraud. Typically, this plan involves the falsification of financial statements. Businesses want to be prepared to combat fraudulent conduct in view of the recent challenging economy (Das, Kumar & Singh, 2022). Financial statement data must be changed in order to generate a fraudulent result before a business administration can report fraud.

Financial statement fraud is planned and criminal conduct done by the top management of publicly traded companies. Three criteria differentiate fraud: rationalisation or attitude, opportunity, and pressure or incentive. The global economy loses millions of dollars each year as a result of several firms failing. It is a sort of management fraud that involves the manipulation of financial information. Numerous organisations place a premium on financial statement fraud prevention (Singh, Singh & Karki, 2021). Deceptive sales reporting is a type of financial reporting fraud. It has made sense in recent years to disclose income in the current year. Expenses should be recorded if they are incurred within the current fiscal year or if they are capitalised in the subsequent fiscal year.

Financial scams have escalated in recent years, and their impact on a variety of stakeholders has prompted researchers and academics to look into the origins of such large-scale frauds. Financial fraud has a negative impact on the economy and the general population; for example, they destroy investors' confidence, resulting in money withdrawal from the capital market (Singh, Singh & Nagpal, 2022). The transition to a digital world over the last decade has opened up new options for conducting business and identifying fraudulent activities, including the use of big data technology, machine-learning fraud detection, data mining, and text mining. The online platform contains a massive amount of unstructured text (Singh, Nagpal & Mundi, 2020). Numerous researchers, businesses, and government agencies are attempting to identify a pattern and track hidden data in order to detect fraud, scandals, and abuses.

The purpose of this review paper is to update the research on financial fraud detection using text-mining techniques. Text mining is an artificial intelligence technique that converts unstructured data to structured data by utilising NLP and machine learning algorithms to enhance the analysis (Table 12.1).

Table 12.1 Topology of Existing Literature

Topology	*Text Mining in HealthCare*	*Text Mining in Retailing and Marketing*	*Text Mining in Finance and Business Intelligence*
Basis and data type	Public health data, patient healthcare statistics	Extraction of knowledge from product line and brand extension to food supply chain networks, customer knowledge is critical	Processes for sharing knowledge in the context of corporate bond classification, textual databases
Type of methodology: data-mining techniques and findings	Classification; clustering, dependency modelling, knowledge management system (KMS)	Dependency modelling; cluster analysis	Classification clustering, hybrid SOFM/LVQ, extension data Mining (EDM), decision tree neural network, multiple key term phrasal knowledge sequences (MKTPKS), group-based knowledge flows (GKFs)
Literature support	Lavrac et al. (2007), Hwang et al. (2008), Luo et al. (2016)	Liao, Chen and Wu (2008), Li et al. (2010)	Cheng, Lu and Sheu (2009), Cantu and Ceballos (2010), Li, Zhu and Pan (2010), Liu and Lai (2011), Ur-Rahman and Harding (2012)

Source: Compiled by Author.

Data mining in the medical field is extremely complex due to the difficulty of locating adequate text data relating to subjects that contain private and confidential information. However, researchers are constantly utilising available resources to conduct biomedical text mining in order to detect fraud and abuse in the healthcare sector.

12.2 Data and Methodology

This study has provided a systematic literature review on the big data literature available on *text mining, Natural Language Processing (NLP), topic modelling, Group method of data handling (GMDH), Multilayer perceptron (MLP), Semantic Analysis, etc.*

Data: We searched a bibliographic database and culled 100 research papers in order to gain a better understanding of text analysis in various domains. We selected 45 papers from these 100 for a review of the role of text mining in detecting fraud and business intelligence. The research findings from the papers that were reviewed have been classified and summarised in Tables 12.2 and 12.3. Table 12.2 summarises a detailed analysis of 15 research papers on financial statement fraud. Table 12.3 details the research that has been conducted in the field of text mining in social media. The distribution is determined by the study's theme, time period, sample data, techniques for analysing large amounts of data, conclusion, and findings.

Objective: The main objective is to address two questions, i.e.,

1. To have a better understanding of the value of data-mining tools in detecting financial fraud.
2. To track the trends of literature in relation to published work related to text mining in financial frauds.

Methodology: The purpose of the aforementioned paper is to ascertain the applicability of data-mining techniques and their trends in the Indian context. There is a wealth of literature available in the form of research papers, working papers, conference proceedings, news, tweets, social media posts, and audio and video content. With regards to the subject, we selected 100 research papers from a reputable online database and thoroughly examined them to determine the theme and appropriate category.

Financial forecasting is also a difficult process to master in the fields of data mining and text mining.

12.3 Text Mining

According to Dorre et al. (1999), text mining and data mining are essentially the same techniques; the only difference is that text is analysed rather than structured data. However, Mathimagal (2018) defines text mining as the process of extracting information in order to obtain a useful pattern.

Text documents are a plentiful and dominant source of relevant information in unstructured form in business, and text mining is an automated technique for analysing various text documents. Financial fraud comes in a variety of forms, including money laundering, false financial statements, forgery of proofs and supporting documents, credit card fraud, securities and commodity fraud, and others. Classification, categorisation, clustering, summarisation, topic identification, and extraction of meaningful structured data are all steps in the text-mining process. It is the examination of words/clusters of words in order to ascertain relationships between various variables.

The papers cited in Table 12.2 indicate that the financial statements themselves indicate about some wrongdoings in an organisation. Detecting financial fraud using advance machine learning techniques can help many innocent stakeholders.

Table 12.3 indicates the role of social media in financial sector. The social media posts available on the Internet track the sentiments of investors, which have a spiralling impact on the capital and performance of the company.

12.4 Conclusion

This chapter is a comprehensive review of the field of financial fraud detection using text mining. Financial statement analysis is the first step in detecting fraud in the financial statements of various businesses. Financial statements of businesses contain both financial and non-financial information. Previously, the emphasis was exclusively on financial data. Through some initial intriguing results, this article demonstrated the critical role of non-financial information in detecting financial fraud. Early detection of fraudulent activity through the use of such data may aid in the prediction of future fraudulent activity, allowing the regulating authorities to take necessary action to avoid such large-scale scams and financial abuses.

Table 12.2 Text Mining a Suitable Approach for Financial Statement Fraud Detection

S. No	*Author*	*Title*	*Journal*	*Data*	*Methodology*	*Findings*
1.	Sun and Li (2006)	Data-mining method for listed companies' financial distress prediction	Elsevier: *Knowledge-Based Systems*	35 financial ratios and 135 pairs of listed companies	• *Data pre-processing algorithm combining AOI and IG* • *Data sorting, particularly of attributes with continuous values*	The authors argued that by combining AOI, IG, and decision trees, they could circumvent those issues and accurately forecast listed companies' financial distress.
2.	Kumar and Ravi (2016)	A survey of the applications of text mining in financial domain	Elsevier: *Knowledge Based Systems*	Analysed 89 papers published between 2000 and 2016	• *An extensive look at the uses of text mining in the financial sector.* • *To split and classify data, use CART. To handle groups of data, use CART (GMDH) -Perceptron multilayer (MLP)*	Text-mining techniques are advantageous for detecting FOREX fraud and stock market forecasting

(*Continued*)

Table 12.2 (*Continued*) Text Mining a Suitable Approach for Financial Statement Fraud Detection

S. No	*Author*	*Title*	*Journal*	*Data*	*Methodology*	*Findings*
3.	Tarjo and Herawati (2015)	Application of Beneish M-score models and data mining to detect financial fraud	*Science Direct: Procedia - Social and Behavioral Sciences*	Financial statements 2001–2014	• *Beneish model M-score* • *Principle component analysis (PCA)* • *Logit regression model*	The Beneish M-Score model can be used to detect financial fraud and the company's prospects
4.	Ibrahim and Ali (2009)	The use of data-mining techniques in detecting fraudulent financial statements: an application on manufacturing firms	*The Journal of Faculty of Economics and Administrative Sciences, Suleyman Demirel University*	Financial statement of 100 Turkish company listed in Istanbul stock exchange	• *Decision tree* • *Neural network*	Auditor-unfamiliar data-mining techniques were used to aid in the detection of financial statement fraud

(*Continued*)

Table 12.2 (*Continued*) Text Mining a Suitable Approach for Financial Statement Fraud Detection

S. No	*Author*	*Title*	*Journal*	*Data*	*Methodology*	*Findings*
5.	Kirkos, Spathis anda Manolopoulos (2007)	Data-mining techniques for the detection of fraudulent financial statements	*Science Direct: Expert Systems with Applications*	Sample data of 76 Greek companies	• *Decision trees* • *Neural networks* • *Bayesian belief network*	Data-mining techniques, which claim to be capable of advanced classification and prediction, may aid auditors in detecting management fraud
6.	Zaki and Theodouildis (2013)	Analysing financial fraud cases using a linguistics-based text-mining approach	PhD research investigated at Manchester Business School, University of Manchester	Comprehensive dataset of manipulation cases occurred in the US stock markets from 1990–2001	• *Text-mining application lexical resources* • *Manipulation participant analyser* • *SEC complain document analysis*	A text-mining approach based on linguistics is an extremely powerful tool for deciphering financial fraud

(*Continued*)

Table 12.2 (*Continued*) Text Mining a Suitable Approach for Financial Statement Fraud Detection

S. No	*Author*	*Title*	*Journal*	*Data*	*Methodology*	*Findings*
7.	Netzer, Feldman, Goldenberg and Fresko (2012)	Mine your own business: market-structure surveillance through text mining	*Marketing Science*	They demonstrated text-mining approach using two cases, i.e., Sedan automobile and diabetes drugs forum	• *Text-mining methodology* • *Semantic network Analysis* • *Sentiment analysis*	The authors emphasised the critical nature of text mining as a marketing tool
8.	Kim, Kang and Jeong (2018)	Text mining and sentiment analysis for predicting box office success	*KsII Transactions on Internet and Information Systems*	233,631 reviews of 147 movies with popularity ratings is collected by a XML crawling package in R programme	• *Classification method on Natural language processing* • *Machine learning algorithm*	Researchers discovered a link between popular demand and box office earnings. It shows that user-generated material is more accurate than standard measurements in determining business performance.

(*Continued*)

Table 12.2 (*Continued*) Text Mining a Suitable Approach for Financial Statement Fraud Detection

S. No	*Author*	*Title*	*Journal*	*Data*	*Methodology*	*Findings*
9.	Bhardwaj, Narayan, Vanraj and Dutta (2015)	Sentiment analysis for Indian stock market prediction using Sensex and nifty	Elsevier: *Procedia Computer Science*	Literature survey	• *Natural language processing* • *Subjective content analysis*	Text mining and advanced python script code can be beneficial in market prediction
10.	Albashrawi (2016)	Detecting financial fraud using data-mining techniques: A decade review from 2004 to 2015	*Journal of Data Science*	The author has examined 65 articles from 2004 to 2015	• *Keywords analysis* • *Subjective content analysis*	Researchers could design projects focusing on regularly used ways by first evaluating how those methods are currently being used and then attempting to implement new methods that rely on comparable settings
11.	Gupta and Gill (2012)	Financial statement fraud detection using text mining	*International Journal of Advanced Computer Science and Applications*	Detailed reviews of nine papers have been conducted	• *Existing academic literature review*	Text-mining methods benefit the extraction of hidden text data

(*Continued*)

Table 12.2 (*Continued*) Text Mining a Suitable Approach for Financial Statement Fraud Detection

S. No	*Author*	*Title*	*Journal*	*Data*	*Methodology*	*Findings*
12.	Gadda and Dey (2014)	Business Intelligence for public sector banks in India: A case study-design, development and deployment	*Journal of Finance, Accounting and Management*	Public and private banks of India	• *Case analysis*	Discussed the need of MIS and BI solutions in India
13.	Gray and Debreceny (2014)	A taxonomy to guide research on the application of data mining to fraud detection in financial statement audits	Elsevier: *International Journal of Accounting Information Systems*	Audited financial statements of companies	• *Investigated fraud –related procedure.* • *Identification of data-mining taxonomy and fraud patterns*	While fraud detection through audit is a safeguard for the company's various stakeholders, auditor fraud is a significant issue that can be addressed using text-mining techniques

(*Continued*)

Table 12.2 (*Continued*) Text Mining a Suitable Approach for Financial Statement Fraud Detection

S. No	*Author*	*Title*	*Journal*	*Data*	*Methodology*	*Findings*
14.	Chen, Liou, Chen and Wu (2019)	Fraud detection for financial statements of business groups	Elsevier: *International Journal of Accounting Information Systems*	Go over the reports, -correspondence with shareholders, and financial news for fraudulent and non-fraudulent assertions of financial information from commercial organisations	• *Algorithm used in filtering fraudulent activities.*	Financial statement analysis aids in the detection of fraudulent activity and can aid in the reduction of investment losses
15.	Sadgali, Sael and Benabbou (2019)	Performance of machine learning techniques in the detection of financial frauds	Science Direct: *Procedia Computer Science*	Taken cases of Accident occurred between the policy issue date and the effective starting date	• *Comparison of Multilayer Feedforward Neural Networks (MLFF), Support Vector Machines (SVM), Genetic Programming (GP), Group Method of Data Handling (GMDH), and Probabilistic Neural Networks (PNN)*	The authors conclude that because each case is unique, hybrid fraud detection techniques tailored to the situation may aid in resolving the real-time problem

Source: Compiled by author.

Table 12.3 Text Mining: An Approach for Detection of Financial Fraud Using Social Media Post

Sl. No	*Author*	*Title*	*Journal*	*Data*	*Methodology*	*Findings*
1.	He, Zha and Li (2013)	Social media competitive analysis and text mining: A case study in the pizza industry	Elsevier: *International Journal of Information Management*	Text content of Facebook and Twitter- of the three largest pizza chains: Pizza Hut, Domino's Pizza and Papa John's Pizza	• *A Case study method* • *Competitive analysis* • *Text mining to analyse unstructured text content*	A competitive analysis utilising text mining is beneficial for comprehending the business pattern. It aids in the acquisition of superior competitive intelligence
2.	Dong, Shaoyi and Zhang (2018)	Leveraging financial social media data for corporate fraud detection	Routledge: *Journal of Management Information Systems*	Database of 64 fraudulent firms and a matched sample of 64 non-fraudulent firms	• *Support Vector Machine (SVM),* • *Neural Networks (NN),* • *Decision Tree (DT), And* • *Logistic Regression (LR)*	Social media is a one-of-a-kind platform for identifying a variety of pertinent information, most notably financial frauds
3.	Goyal, Singh and Sharma (2020)	Fraud detection on social media using data analytics	*International Journal of Engineering Research & Technology*	News: low quality bogus news and tweets	• *Entity link analysis using graph database* • *Robust fraud detection by social network analysis (SNA)* • *Neural networks*	It is critical to identify fake news via data analysis in order to combat financial fraud

(*Continued*)

Table 12.3 (*Continued*) Text Mining: An Approach for Detection of Financial Fraud Using Social Media Post

Sl. No	*Author*	*Title*	*Journal*	*Data*	*Methodology*	*Findings*
4.	Mathimagal (2018)	Social network analytics (SNA) fraud	*International Journal of Pure and Applied Mathematics*		• *Social network analytics*	Large-scale social networks contain information that is dispersed across a heterogeneous network. Machine learning is advantageous for detecting and forecasting these types of frauds
5.	Zhoa (2013)	Analysing Twitter data with text mining and social network analysis	*11th Australasian Data-Mining Conference (AusDM 2013), Canberra, Australia, November 2013*	Official Department of Immigration and Citizenship (DIAC) Twitter accounts	• *Social network analysis of twitter followers and retweeting*	Both data-mining and text-mining techniques are beneficial for analysing tweet networks and their impact on a large population

Source: Compiled by author.

References

Albashrawi, M. (2016). Detecting financial fraud using data mining techniques: a decade review from 2004 to 2015. *Journal of Data Science*, 14(3), 553–569.

Bach, M., Krstić, Ž., Seljan, S., & Turulja, L. (2019). Text mining for Big Data Analysis in financial sector: a literature review. *Sustainability Review*, 11(5), 1277.

Bagale, G.S., Vandadi, V.R., Singh, D. et al. (2021). Small and medium-sized enterprises contribution in digital technology. *Annals of Operations Research*. Doi: 10.1007/s10479-021-04235-5.

Cantú, F.J. & Ceballos, H.G. (2010). A multiagent knowledge and information network approach for managing research assets. *Expert Systems with Applications*, 37(7), 5272–5284.

Chen, F., Chi, D., Zhu, J. (2014). Application of random forest, rough Set theory, decision tree and neural network to detect financial statement fraud –taking corporate governance into consideration, In D.-S. Huang, V. Bevilacqua, P. Premaratne (Eds.), *10th International Conference Intelligent Computing Theory*, ICIC 2014, Springer, Taiyuan, 221–234.

Chen, Y. J., Liou, W. C., Chen, Y. M., & Wu, J. H. (2019). Fraud detection for financial statements of business groups. *International Journal of Accounting Information Systems*, 32, 1–23.

Cheng, H., Lu, Y.-C, Sheu, C. (2009). An ontology-based business intelligence application in a financial knowledge management system. *Expert Systems with Applications*, 36(2), 3614–3622.

Das, U., Kumar, S., & Singh, A. (2022). Gender, technology and innovation: the role of women in Indian micro, small and medium enterprises. World Review of Entrepreneurship Management and Sustainable Development, 18(4), 429. https://doi.org/10.1504/wremsd.2022.10046819

Dorre, J., Gerstl, P., Seiffert, R. (1999). Text mining: Finding nuggets in mountains of textual data. In *Proceedings of 5th ACM International Conference on Knowledge Discovery and Data Mining (KDD-99)*, San Diego, CA, ACM Press, 398–401.

Gadda, K. R., & Dey, S. (2014). Business Intelligence for Public Sector Banks in India: A Case study-Design, Development and Deployment. *Journal of Finance, Accounting & Management*, 5(2).

Gray, G.L., Debreceny, R.S. (2014). A taxonomy to guide research on the application of data mining to fraud detection in financial statement audits. *International Journal of Accounting Information Systems*, 15, 357–380.

Gupta, R., & Gill, N. S. (2012). Financial statement fraud detection using text mining. *Editorial Preface*, 3(12), 189–191.

Hwang, H.G., Chang, I.C., Chen, F.J. & Wu, S.Y. (2008). Investigation of the application of KMS for diseases classifications: A study in a Taiwanese hospital. *Expert Systems with Applications*, 34(1), 725–733.

Ibrahim, H., Ali, H. (2009) The use of data mining techniques in detecting fraudulent financial statements: An application on manufacturing firms. *The Journal of Faculty of Economics and Administrative Sciences*, 14(2), 157–170.

Kautish, S. (2008). Online banking: A paradigm shift. *E-Business, ICFAI Publication, Hyderabad*, 9(10), 54–59.

Kautish, S., Singh, D., Polkowski, Z., Mayura, A. & Jeyanthi, M. (2021). *Knowledge Management and Web 3.0: Next Generation Business Models.* De Gruyter, Berlin.

Kautish, S., Thapliyal, M.P. (2013). Design of new architecture for model management systems using knowledge sharing concept. *International Journal of Computer Applications*, 62(11).

Kim, Y., Kang, M., & Jeong, S. R. (2018). Text mining and sentiment analysis for predicting box office success. *KSII Transactions on Internet and Information Systems (TIIS)*, 12(8), 4090-4102.Bhardwaj, Narayan, Vanraj and Dutta (2015),

Kirkos, E., Spathis, C., Manolopoulos, Y. (2007). Data mining techniques for the detection of fraudulent financial statements. *Expert Systems with Applications*, 32, 995–1003.

Kumar B, S and Ravi, V. (2016). A survey of the applications of text mining in financial domain, *Knowl.-Based Syst.* 114 128–147

Lavrač, N., Bohanec, M., Pur, A., Cestnik, B., Debeljak, M., & Kobler, A. (2007). Data mining and visualization for decision support and modeling of Public Health-care resources. *Journal of Biomedical Informatics*, 40(4), 438–447.

Li, N., Liang, X., Li, X., Wang, C., Wu, D.D. (2009). Network environment and financial risk using machine learning and sentiment analysis. *Human Ecol Risk Assess Int J*, 15(2), 227–252.

Li, X., Zhu, Z. & Pan, X. (2010). Knowledge cultivating for intelligent decision making in small & middle businesses. *Procedia Computer Science*, 1(1), 2479–2488.

Liao, S.-H., Chen, C.-M., & Wu, C.-H. (2008). Mining customer knowledge for product line and brand extension in retailing. *Expert Systems with Applications*, 34(3), 1763–1776.

Liu, D.R. & Lai, C.H. (2011). Mining group-based knowledge flows for sharing task knowledge. *Decision Support Systems*, 50(2), 370–386.

Luo, E., Bhuiyan, M. Z., Wang, G., Rahman, M. A., Wu, J., & Atiquzzaman, M. (2018). Privacyprotector: Privacy-protected patient data collection in IOT-based Healthcare Systems. *IEEE Communications Magazine*, 56(2), 163–168.

Mathimagal M, N. (2018). Social Network Analytics (Sna) Fraud. *International Journal of Pure and Applied Mathematics*, 118(20), 191–202.

Netzer, O., Feldman, R., Goldenberg, J., & Fresko, M. (2012). Mine your own business: Market-structure surveillance through text mining. *Marketing Science*, 31(3), 521–543.

Sadgali, I., Sael, N., & Benabbou, F. (2019). Performance of machine learning techniques in the detection of financial frauds. *Procedia Computer Science*, 148, 45–54.

Singh, A., Gite, P. (2015). Corporate governance disclosure practices: A comparative study of selected public and private life insurance companies in India. *Apeejay - Journal of Management Sciences and Technology*, 2(2).

Singh A., Singh H., Singh A. (2022). People analytics: Augmenting horizon from predictive analytics to prescriptive analytics. In: Jeyanthi P.M., Choudhury T., Hack-Polay D., Singh T.P., Abujar S. (eds). *Decision Intelligence Analytics and the Implementation of Strategic Business Management.* EAI/Springer Innovations in Communication and Computing. Springer, Cham. https://doi.org/10.1007/978-3-030-82763-2_13

Singh H., Singh A., Nagpal E. (2022). Demystifying behavioral biases of traders using machine learning. In: Jeyanthi P.M., Choudhury T., Hack-Polay D., Singh T.P., Abujar S. (eds). *Decision Intelligence Analytics and the Implementation of Strategic Business Management.* EAI/Springer Innovations in Communication and Computing. Springer, Cham. https://doi.org/10.1007/978-3-030-82763-2_16

Singh, Anurag, Nagpal, E., & Mundi, H. S. (2020). Brand personification through celebrity ambassador: A study to investigate the impact on consumer attitude and loyalty. International Journal of Business Excellence, 1(1), 1. https://doi.org/10.1504/ijbex.2020.10034734

Singh, D., Singh, A., Karki, S. (2021). Knowledge management and Web 3.0: Introduction to future and challenges. In *Knowledge Management and Web 3.0.* De Gruyter, Cambridge University Press. Doi: 10.1515/9783110722789-001Agents.

Sun, J., Li, H. (2006). Data mining method for listed companies' financial distress prediction. *Knowledge-Based Systems*, 21(1), 1–5.

Ur-Rahman, N. & Harding, J.A. (2012). Textual data mining for industrial knowledge management and text classification: A business oriented approach. *Expert Systems with Applications*, 39, 4729–4739.

Zaki, M., and Theodoulidis (2013), Analyzing Financial Fraud Cases Using a Linguistics-Based Text Mining Approach, (PhD Research) Manchester Business School, University of Manchester

Chapter 13

Effectiveness of Robotics Process Automation in Increasing the Productivity of Employees and Organizations with Reference to Business Analytics

Anni Arnav
Presidency University

Vishal Srivastava
JAIN (Deemed-to-be University)

K. R. Varsha
Dayananda Sagar University

Contents

DOI: 10.4324/9781003184928-13

13.1 Background and Introduction

Business intelligence and analytics have paved the way for looking at restructuring the processes and businesses for strategic decision-making and business sustainability in the current business scenario. The timing is such where the leaders in organizations are looking toward transforming the business with the effective utilization of Business Analytical tools in decision making for increased focus on sustainability and cost reduction factor. Thanks to technologies, such as Artificial Intelligence, Internet of Things, Robotics Process Automations (RPAs), and Business Analytics with reference to different domains and functionalities of the business in the organizations. Automations have replaced the employees from doing the repetitive and mundane tasks. The same set of employees can be used where we have to really look at human interventions.

13.2 Problem Context

The study aims at understanding the impact of Robotics Process Automation (RPA) at the workplace related to the HR activities in the organization. Organizations are increasingly focused toward enhancing productivity through various analytical tools and processes in business decisions. There is a need to understand how RPA will affect the overall performance of an organization with respect to cost saving, upgradation of technology, acceptance of change, maintenance of accuracy, and the results drawn out of the same. Researchers have tried to analyze based on the responses from the population working on the RPA technology and find the effectiveness and implications of the same.

13.3 Objectives

- To study the effectiveness of RPA on the productivity of the organization and analytical business decisions.
- To identify the effect of RPA on employees in increasing the productivity of the organization and effective managerial decisions.

13.4 Literature Review

Automation is not a new idea and has been a gift for more than a century where people usually have been looking to grow the performance of work strategies and used one-of-a-kind tools to gain this (Kautish, 2008, Kautish & Thapliyal, 2013). It has been a consistent dialogue over the past decades, of whether automation and new technology will and should replace sure jobs absolutely or simply act as assist (Nawaz, 2019). The emergence of technology puts work responsibilities, which are achieved by using people, who've not been threatened to get replaced by way of generation before, in danger of being partially or completely automatic, ultimately endangering positive activity in job classifications (Delen & Ram, 2018). But, despite the fact that automation has put certain job categories and work duties at risk of decreasing or being eliminated, it has additionally created new work tasks and categories (Asatiani & Penttinen, 2016). A top-level view, the employee participation in the context of undertaking process enrichment, includes enhancement of the worker's process pride and overall performance via allowing them to take part in the managerial choices (Issac, Muni & Desai, 2018). Work responsibilities being monotonous and repetitive, RPA can replace this type of duties to permit the employee with the awareness of unstructured work tasks, which cannot absolutely be performed with the aid of robot technique automation itself and calls for human involvement (Gambao, et al., 1997). RPA is used to automate the work tasks of services (Issac et al., 2018). As a way to be an employee, the employee desires to have a job that requires positive talents (Singh & Gite, 2015). The required abilities depend on the work obligations (Bagale et al., 2021). Moreover, whilst the employee makes use of RPA as a support to automate structured work responsibilities, this will generate in the employee perceiving consequences on their task (Davenport, 2018). These perceived results may be; activity delight, job expansion, job enrichment, job satisfaction, upskilling and deskilling (Anagnoste, 2018). There is an argument of the objective behind automation and replacement of human workforce (Rai, et al., 2019). RPA is not a robot but it is being used as software to solve the problems (De Silva, 1997). The writer argues that automation frequently intends to replace the workforce (Kautish, Singh, Polkowski, Mayura & Jeyanthi, 2021). But, he further discusses that automation additionally complements personnel, which could be very much like the case for RPA. In addition, A report from 2015 located the number one goal in the back of the selection to apply RPA in the place of employees (Sharma, 2016). Several applications areas can be developed to make jobs automated (Gupta, 2009). The author contends that the undertaking attributes employee participation and aim internalization ought to both bring about employees experiencing an enhancement of the process characteristic "task identity," when you consider that these attributes may want to boom the extent of impact the personnel understand that their work mission has at the entire work technique (Pan, 2010). In order to increase the production efficiency, the automation process started with tools to replace the humans (Willcocks et al., 2015). RPA helps in managing unstructured tasks with more complex problems (Devarajan, 2018).

According to researchers, RPA is the next big thing in outsourcing industry. So based on their study, we see that they emphasize on changing from low-skilled cheap human resources to process automation, which helps in cost reduction, also by removing human error, which in turn helps in better optimization of process automation, which results in decreasing the cycle time. RPA is a challenging task to implement because of the existing IT processes & it is usually done by local business units. So, based on the report, we understand the organizational consequences of having local business units manage RPA initiatives, hence should be implemented in the right hands. The study focuses on the RPA and employee at individual level (Bahrin et al., 2016).

The article talks about the complex systems and the controlling factors are the most difficult tasks in the current scenario with the advent of technology. There is a necessity of finding out the essence of it with a thorough understanding. Organizations are more inclined towards the usage and capacity building with the latest analytical competencies to be more competitive in the organizational strategy. Companies look at predicting the future problems related to the demand and supply aspects in the short and long term through analytics (Davenport & Harris, 2017). The organizations are looking at a major shift in adapting to the analytics talent, which is required for the expansion plans (Franks, 2014). The book focuses on the importance of predictive analytics and its effect on day-to-day life. The author further states the reasons for predicting human behaviour to enhance the sales perspective (Siegel, 2016). The study highlights the different factors of business analytics in organizations (Rachida, 2017).

Researchers have found usage of HR Analytics to enhance organizational performance is very low. HR Analytics has the capacity to bring in a big difference to the value of HR leaders in making appropriate decisions in organizations with the addition of intuition. HR Analytics generates an increased return on investments (ROI) with respect to HR tasks (Ben-Gal, 2019). Analytics-based approach is very much beneficial in identifying high potential leaders with the evaluation of certain competencies, which are very important for business decisions (Mondare & Carson, 2011).

13.5 Methodology

The study was conducted in selected IT and ITES employees of Bengaluru City. 100 samples were collected using non-probabilistic convenience sampling techniques. Pilot study was conducted for 40 samples to ensure that the purpose of design of the study is related with the research objectives. Structured questionnaires were framed and collected data was analyzed using SPSS 20.0.

13.6 Hypothesis

H1: There is an association between Educational Qualification and RPA, which can help fill the shortage of skilled workers that is achieved from RPA.

H2: There is an association between the understanding of the real meaning of RPA and RPA can save time on repetitive tasks that are achieved from RPA.

H3: There is an association between the understanding of the real meaning of RPA, which can significantly reduce costs that is achieved from RPA.

H4: There is an association between the understanding of the real meaning of RPA and RPA implementation costs too high compared to ROI that is achieved from RPA.

H5: There is a significant relationship between RPA and improved efficiency of the organization.

13.7 Data Analysis

13.7.1 Respondents Based on Age Working on RPA

Inference: A majority of the responses fall under the age group between 20 and 25 years, which indicates that the younger population works on RPA and are updated on the current technology. The least responses are from the age group more than 40 years, which are barely updated on the current technology (Table 13.1).

13.7.2 Respondents Based on Gender Working on RPA

Inference: A majority of the responses are from the female candidates working on RPA, which indicates that the major usage of the RPA is done by the female responses. But, there is a little difference in the percentage of male and female responses with the least responses coming from male candidates (Table 13.2).

Table 13.1 Respondents Based on the Age Working on RPA

Age Group					
		Frequency	*Percent*	*Valid Percent*	*Cumulative Percent*
Valid	20–25 years	36	36.0	36.0	36.0
	26–30 years	18	18.0	18.0	54.0
	31–35 years	18	18.0	18.0	72.0
	35–40 years	17	17.0	17.0	89.0
	> 40 years	11	11.0	11.0	100.0
	Total	100	100.0	100.0	

Table 13.2 Respondents Based on the Gender Working on RPA

Gender					
		Frequency	*Percent*	*Valid Percent*	*Cumulative Percent*
Valid	Male	49	49.0	49.0	49.0
	Female	51	51.0	51.0	100.0
	Total	100	100.0	100.0	

13.7.3 KMO and Bartlett's Test

The value 0.767 shows that sampling is adequate for analysis.

The alpha coefficient for the sixteen items is 0.727, suggesting that the items have relatively high internal consistency (Tables 13.3 and 13.4).

13.7.4 Chi Square Analysis

Relationship between Educational Qualification and RPA can help fill the shortage of skilled workers that is achieved from RPA.

Relationship between the Educational Qualification and RPA can help fill the shortage of skilled workers that is achieved from RPA.

Inference: Chi-square test was conducted and the results summarized in Table 13.5 clearly have shown that the Pearson Chi-square value is less than 0.05, indicating that null hypothesis is rejected.

Table 13.3 KMO and Bartlett's Test

KMO and Bartlett's Test		
Kaiser-Meyer-Olkin measure of sampling adequacy		0.767
Bartlett's test of sphericity	Approx. chi-square	590.279
	Df	120
	Sig.	0.000

Table 13.4 Reliability Statistics

Cronbach's Alpha	*No of Items*
0.727	16

Table 13.5 Relationship between the Educational Qualification and Robotic Process Automation Using Chi-Square Tests: Shortage of Skilled Workers

Chi-Square Tests			
	Value	*Df*	*Asymp.iSig. (2-sided)*
Pearson chi-square	22.406[a]	12	0.033
Likelihood ratio	23.671	12	0.023
Linear-by-linear association	0.048	1	0.827
No of valid cases	100		

[a] 11 cells (55.0%) have expected count less than 5. The minimum expected count is.91.

13.7.5 Relationship between the Understanding of the Real Meaning of RPA and RPA Can Save Time on Repetitive Tasks

Relationship between the understanding of the real meaning of RPA and RPA can save time on repetitive tasks.

Inference: Chi-square test was conducted and the results from Table 13.6 clearly shows that the Pearson Chi-square value is less than 0.05, indicating that null hypothesis is rejected.

Table 13.6 Relationship between the Understanding of the Real Meaning of Robotic Process Automation and Robotic Process Automation Using Chi-Sqaure Tests: Can Save Time on Repetitive Tasks

Chi-Square Tests			
	Value	*Df*	*Asymp.iSig. (2-sided)*
Pearson chi-square	45.411[a]	16	0.000
Likelihood ratio	34.010	16	0.005
Linear-by-linear association	8.469	1	0.004
No of valid cases	100		

[a] 19 cells (76.0%) have expected count less than 5. The minimum expected count is 14.

13.7.6 Relationship between the Understanding of the Real Meaning of Robotic Process Automation and RPA Can Significantly Reduce Costs

Relationship between the understanding of the real meaning of RPA and RPA can significantly reduce costs.

Inference: Chi-square test was conducted and the results from Table 13.7 clearly have shown that the Pearson Chi-square value is less than 0.05, indicating that the null hypothesis is rejected.

13.7.7 Relation between the Understanding of the Real Meaning of RPA and RPA Implementation Costs Too High Compared to ROI

Relation between the understanding of the real meaning of RPA and RPA implementation costs is too high compared to ROI.

Inference: Chi-square test was conducted and the results from Table 13.8 clearly have shown that the Pearson Chi-square value is less than 0.05, indicating that the null hypothesis is rejected.

Regression

Linear Regression

H0: There is no significant relationship between RPA and improved efficiency of the organization.

H1: There is a significant relationship between RPA and improved efficiency of the organization.

Table 13.7 Relationship between the Understanding of the Real Meaning of Robotic Process Automation and Robotic Process Automation Using Chi-Square Tests: Reduce Costs

Chi-Square Tests			
	Value	*df*	*Asymp.iSig. (2-sided)*
Pearson chi-square	29.579[a]	16	0.020
Likelihood ratio	31.946	16	0.010
Linear-by-linear association	2.937	1	0.087
No of valid cases	100		

[a] 16 cells (64.0%) have expected count less than 5. The minimum expected count is 28.

Table 13.8 Relation between the Understanding of the Real Meaning of Robotic Process Automation and RPA Implementation: Costs Too High

Chi-Square Tests			
	Value	*df*	*Asymp.iSig. (2-sided)*
Pearson chi-square	26.643[a]	16	0.046
Likelihood ratio	24.025	16	0.089
Linear-by-linear association	0.644	1	0.422
No of valid cases	100		

[a] 18 cells (72.0%) have expected count less than 5. The minimum expected count is 56.

Table 13.9 Model Summary

Model Summary[b]					
Model	*R*	*R^2*	*Adjusted R^2*	*Std. Error of the Estimate*	*Durbin-Watson*
1	0.521[a]	0.271[b]	0.264	0.69262	1.961

[a] Predictors: (Constant), RPA.
[b] Dependent variable: improved efficiency.

Here, the R-square is 0.271, representing that independent variables explain 27.1% of the variability of our dependent variable. The Durbin Watson value is 1.961, which is closer to 2 signifying no autocorrelation in the dataset (Table 13.9).

13.7.8 ANOVA

The *F*-ratio in the **ANOVA** Table 13.10 tests whether the overall regression model is a good fit for the data. Here, this table depicts the model being fit for the data (Table 13.11).

Here, the linear regression equation can be represented as =

$$Y = a + B X$$

Y=Improved Efficiency (Dependent Variable)
a=Constant (1.116)
B=Unstandardized Coefficients (0.571)
X=RAP (Independent Variable)

Table 13.10 ANOVA

ANOVA[a]						
Model		*Sum of Squares*	*df*	*Mean Square*	*F*	*Sig.*
1	Regression	17.515	1	17.515	36.512	0.000[b]
	Residual	47.012	98	0.480		
	Total	64.528	99			

[a] Dependent variable: improved efficiency.
[b] Predictors: (Constant), RPA.

Table 13.11 Coefficients

Coefficients[a]						
		Unstandardized Coefficients		*Standardized Coefficients*		
Model		*Std. Error*	*Beta*		*t*	*Sig.*
1	(Constant)	1.116	0.231		4.822	0.000
	RAP	0.571	0.094	0.521	6.042	0.000

[a] Dependent variable: improved efficiency.

The regression equation as depicted in the table clearly shows that RAP has a significant relationship with Improved Efficiency and impacts it by 0.571 times, which is an important contribution. Hence, the RAP is directly related to the improved efficiency of the organization. The P-value is less than 0.05, hence rejecting the null hypothesis and accepting the alternate hypothesis.

13.8 Findings and Discussion

The study indicates that the younger population works on RPA and are updated on the current technology. The sample population who have studied post-graduation infer that the respondents are better qualified who are allowed to work on RPA since it requires better experience and knowledge about the process and implementation of the same. The salaried class infers that the organization that is doing good prefers to invest on trending technology, which is the RPA and the employees working under those types of organizations get to work on RPA and make difference in the system. It also makes the management take business decisions appropriately. Chi-square test was conducted and the results clearly have shown that the Pearson

Chi-square value is less than 0.05, indicating that there is an association between understanding the real meaning of RPA and it can save time on repetitive tasks. RPA is directly related to the improved efficiency of the organization. RPA can help fill the shortage of skilled workers as per the findings.

13.9 Recommendations

More respondents between the age group of 26–30 years can be assigned to work on RPA, which would decrease cost and increase the chances of effective business decisions. A proper platform should be provided to them to learn and implement the same in their business. It is recommended to upgrade the responses according to their qualification and tell them how important it is to fill the gaps in an organization using RPA through analytical reports. Once the person working on RPA understands the exact meaning of it then he or she should be taking measures to reduce cost. ROI should be always higher when compared to the money spent on the implementation of RPA for which the cost budget and other related costs should be taken care accordingly through the analytical reports.

13.10 Conclusion

RPA is an emerging technology that is being implemented at every step of the business to make the work process easier and try to get more effective results out of the same. But while implementing the same there are several challenges that could be encountered which becomes a hurdle for most of the companies. Sometimes it could be the cost factor or acceptance of the technology by the employees and making it effective or could be worrying for laying back or removing certain employees from the organization.

Therefore, this study was done to understand the risks involved in the RPA process, how the employees think about the process, what is the acceptance level of the process, what type of employees work on them, how it can be improved by resolving all the odds, how does Business Analytics, HR Analytics play a dominant role in enhancing the effect and impact on business decisions. Hence, to conclude, it is important first to understand the right meaning of the process, why we are doing it, how we are doing it and what is the purpose of implementing it. Later, we will be able to analyze the risks and take measures to reduce them and become successful.

References

Anagnoste, S. (2018). Setting up a robotic process automation center of excellence. *Management Dynamics in the Knowledge Economy*, 6(2), 307–332.

Asatiani, A., & Penttinen, E. (2016). Turning robotic process automation into commercial success–case opuscapita. *Journal of Information Technology Teaching Cases*, 6(2), 67–74.

Bagale, G.S., Vandadi, V.R., Singh, D. et al. (2021). Small and medium-sized enterprises' contribution in digital technology. *Annals of Operations Research*. Doi: 10.1007/s10479-021-04235-5.

Bahrin, M.A.K., Othman, M.F., Azli, N.N., & Talib, M.F. (2016). Industry 4.0: A review on industrial automation and robotics. *Jurnalteknologi*, 78(6–13), 137–143.

Ben-Gal, H.C. (2019). An ROI-based review of HR analytics: Practical implementation tools. *Personnel Review*, 48(6), 1429–1448.

Davenport, T.H. (2018). From analytics to artificial intelligence. *Journal of Business Analytics*, 1(2), 73–80.

Davenport, T.H., & Harris, J.G. (2017). *Competing in Analytics*. Boston, MA: Harvard Business Review Press.

De Silva, C.W. (1997). Intelligent control of robotic systems with application in industrial processes. *Robotics and Autonomous Systems*, 21(3), 221–237.

Delen, D., & Ram, S. (2018). Research challenges and opportunities in business analytics. *Journal of Business Analytics*, 1(1), 2–12.

Devarajan, Y. (2018). A study of robotic process automation use cases today for tomorrow's business. *International Journal of Computer Techniques*, 5(6), 12–18.

Franks, B. (2014). *The Analytics Revolution: How to Improve Your Business by Making Analytics Operational in the Big Data Era*. New York: Wiley.

Gambao, E., Balaguer, C., Barrientos, A., Saltaren, R., & Puente, E.A. (1997). Robot assembly system for the construction process automation. In *Proceedings of International Conference on Robotics and Automation* (vol. 1, pp. 46–51). IEEE.

Gupta, A.K., & Arora, S.K. (2009). *Industrial Automation and Robotics*. Laxmi Publications.

Issac, R., Muni, R., & Desai, K. (2018). Delineated analysis of robotic process automation tools. In *2018 Second International Conference on Advances in Electronics, Computers and Communications (ICAECC)* (pp. 1–5). IEEE.

Kautish, S. (2008). Online banking: A paradigm shift. *E-Business, ICFAI Publication, Hyderabad*, 9(10), 54–59.

Kautish, S., Singh, D., Polkowski, Z., Mayura, A. & Jeyanthi, M. (2021). *Knowledge Management and Web 3.0: Next Generation Business Models*. Berlin: De Gruyter.

Kautish, S., & Thapliyal, M.P. (2013). Design of new architecture for model management systems using knowledge sharing concept. *International Journal of Computer Applications*, 62(11).

Mondare, S., Douthitt, S., & Carson, M. (2011). Maximizing the impact and effectiveness of HR Analytics to drive business outcomes. *People & Strategy*, 34, 20–27.

Nawaz, d.N. (2019). Robotic process automation for recruitment process. *International Journal of Mechanical Engineering and Technology*, 10(3).

Pan, Z., Polden, J., Larkin, N., Van Duin, S., & Norrish, J. (2010). Recent progress on programming methods for industrial robots. In *41st International Symposium on robotics) and Robotik 2010 (6th German Conference on Robotics)*, pp. 1–8.

Parks, R.F. & Thambusamy, R. (2017). Understanding business analytics success and impact: A qualitative study. *Information Systems Education Journal (ISEDJ)*, 15(6), 43.

Rai, D., Siddiqui, S., Pawar, M., & Goyal, S. (2019). Robotic process automation: The virtual workforce. *International Journal on Future Revolution in Computer Science & Communication Engineering*, 5(2), 28–32.

Sharma, k.L.S. (2016). *Overview of Industrial Process Automation.* Elsevier.

Siegel, E. (2016). *Predictive Analytics.* New York: Wiley.

Singh, A., & Gite, P. (2015). Corporate governance disclosure practices: A comparative study of selected public and private life insurance companies in India. *Apeejay - Journal of Management Sciences and Technology* 2(2).

Singh, D., Singh, A., & Karki, S. (2021). Knowledge management and web 3.0: Introduction to future and challenges. In *Knowledge Management and Web 3.0: Next Generation Business Models*, De Gruyter, 1–14. Doi: 10.1515/9783110722789-001.

Willcocks, L.P., Lacity, M., & Craig, A. (2015). The IT function and robotic process automation.

Chapter 14

Business Intelligence Application through Customer Relationship Management in LIC of India: A Case Study

Furquan Uddin
Aliah University

Contents

14.1 Introduction

In present times and future, the key factors that will control, expand, and endeavour any business are the interpretation of data available to the organisations (Rygielski et al., 2002). Beyond any geographical boundaries or even beyond the limitations of this planet, organisations will strive to make their mark amongst humankind through the weapons of data and information. As a matter of fact, the quality and quantity of data management will be

DOI: 10.4324/9781003184928-14

of paramount significance to the corporate world (Athanassopoulos, 1997). Nowadays, organisations own an enormous volume of data (Manogaran & Lopez, 2017; Chen & Zhang, 2014). Because of the poor quality of enormous amounts of data, whether useful or not, big investments are being made in information technology by every firm (Williams & Williams, 2006).

Business Intelligence (BI), an innovative approach, explores data and initiates considerable data to support business supervisors, corporate administrators, and diverse set of clients (Sauter, 2014; Popovič et al., 2012). It is useful to provide a huge amount of quality data, which can be helpful in correct and timely decision-making (Williams & Williams, 2006; Yeoh & Popovič, 2016). Concomitantly, most big organisations in the world have largely adopted BI systems (Yeoh & Koronios, 2010; Alshawi et al., 2011). Several studies have been conducted to determine the factors broadly responsible for successful BI implementation (Yeoh & Popovič, 2016). It is also evidenced through several studies that certain dimensions, such as long-term strategy, user participation, approach of top-level management, change management, and organisational resources, might have a strong impact (Kautish, Singh, Polkowski, Mayura & Jeyanthi, 2021). On the contrary, it wasn't unanimous which particular factors were responsible for success (Yeoh & Popovič, 2016; Dooley et al., 2018; Villamarín & Diaz Pinzon, 2017; Nasab et al., 2017).

Almost any business process can benefit from BI, which provides a holistic view and enables teams to look at their own data to find efficiencies and make good daily decisions (Kautish, 2008, Kautish & Thapliyal, 2013). Data investments can help companies accomplish better results thanks to the innovations made in their BI tools (Bagale et al., 2021). Digital transformation has now been deemed a key strategic initiative. Modern BI platforms, which support data access, interactivity, analysis, discovery, sharing, and governance, have arisen in response to the increased demand for the knowledge these platforms make available. This chapter will highlight major companies that have made a BI platform successful through their work with it. Here are five actual, on-the-ground examples of BI platforms in action. BI application tools entail Data Mining, Data Warehouse, Decision Support System, Extract Transformation Loading, Supply Chain Management, Customer Relationship Management (CRM), Business Process Management, Artificial Intelligence, Enterprise Resource Planning, Quality Management System, and Strategic Management (Olszak & Ziemba, 2007; Vercellis, 2009; Arunachalam & Kawalek, 2018). The software's data warehouses or data marts are enormous, and they use data mining, digging, and other activities to retrieve relevant findings and reports. A wide range of operations encompasses BI, including statistical analysis, querying and reporting, business performance analysis, benchmarking, online analytical processing, decision support systems, forecasting, and predictive analysis. It gives businesses information that is beneficial to business planning, such as information about employees, customers, suppliers, and other business associates, and may therefore be used to help with the decision-making process. CRM is one of the viable and profitable strategic tools (Chalmeta, 2006; Zineldin, 2005), which is used across the industries in general and life insurance in particular. In India, there are 24 life insurance firms, and amongst these, LIC of India is the only public entity and market leader (Siddiqui, 2020; Chandrapal, 2019).

The LIC of India has played a crucial role in the socioeconomic development of India since its establishment in 1956. The performance of LIC has been commendable from several perspectives, such as branch expansion, new-business premium customer base, and agency network, and has a noteworthy role in accelerating life insurance extensively across the country (Singh, Singh & Karki, 2021). In tune with the objectives of nationalisation, for the benefit and development of the community at large, the LIC has mobilised the life insurance funds invested by people. The vision of LIC is to emerge as a world class-centric organisation. In view of the BI application through CRM, this case study is a humble attempt to look at the CRM practices in the LIC of India.

14.2 Literature Review

Most of the studies pertaining to BI have been instigated by developed countries, such as the United States, Canada, Japan, and Western European countries while developing countries are also in the queue of BI issues but in a limited range (Acheampong & Moyaid, 2016; Bakunzibake et al., 2016; Hatta et al., 2017; Owusu et al., 2017). BI and CRM have been studied in several areas. However, from the viewpoint of this study, there are some studies presented here. Al-Zadjali and Busaidi (2018) examined the effects of BI on CRM in the telecommunication sector, Oman, and reported mixed impacts, such as there is a positive impact on business process values, and customer values by implementing BI in marketing and BI in customer service has a positive effect on employees' values. Bhat and Darzi (2016) developed a model to explain the effects of CRM dimensions, such as customer knowledge, complaint resolution, customer empowerment, and customer orientation on customer loyalty and competitive advantage of a bank, and found positive impacts. Uddin (2020) highlighted the significance of CRM in life insurance business and the steps involved in applying the CRM process to retain existing policyholders. Kannan and Vikkraman (2015) studied 42 variables in the implementation of CRM processes by public and private life insurers (Singh & Gite, 2015). They found significant differences between both the players. The private players focused on redressal machinery and personalisation, whereas public players concentrated on warehousing, claim payment security, and data mining in the context to CRM process implementation. Further, they reported that private players did better than public players in terms of redressal machinery and claim payment security. Rani (2012) analysed that customers could not make main transactions online without getting in touch with the company in person. As a matter of fact, all companies are not fully integrated with IT in the CRM process. Besides, many insurance companies have a broad database of their policyholders with the support of information technology.

In view of the above literature reviews, BI and CRM have been well researched and comparative studies have also been done in the life insurance sector; however, there is a lack of studies pertaining to the application of BI, particularly CRM practices by a life insurer (LIC of India). This study makes an attempt to fill this research gap, and this is unique in nature as it brings the practices of the organisation based on personal interviews. This study will be of great concern to other life insurers across the world to understand the CRM application.

14.3 Methodology

As per the objective of the study, LIC of India, a market leader, has been selected among 24 life insurance companies because of its strong customer base in India in terms of customer acquisition as well as customer retention. Thereafter, the divisional office located in Varanasi, India, has been targeted for the purpose of primary data collection. A personal interview of the CRM Manager, LIC of India, as a method of eliciting data has been used. The questions asked were on how the CRM process takes place in the organisation. In other words, the process, elements, employees' roles, subunits' roles, overall execution, and the expected outcome were questioned while interviewing the concerned person. The qualitative data collated has been presented through narrative analysis.

14.4 Analysis and Findings

Based on the data collected, the following processes and activities have been in application by LIC of India:

- **CRM Model**: Under the CRM Model, prescribed elements and results are the input and the output. The gap between input and output is accomplished by the CRM process (Figure 14.1).
- **Input (Prescribed Elements) – Customer Contact and Feedback Processing**: Structured customer contacts are essential in CRM and it is the job of the CRM manager to organise various customer contact programmes, at the local level as well as under corporate instructions. This can be either at the branch level or at the divisional level. Such contacts shall be focused on three things: (1) customer education; (2) relationship renewal; and (3) collection, recording, and escalation of information about customer's needs and feelings.
- **Redressal of Grievances**: CRM manager is the single point of contact for the redressal of all customer grievances within specified benchmarks laid down at the corporate level.
- **Special Structures – ECS Cell**: Electronic Clearing Service (ECS), an alternate channel of payment premium, managed at divisional level, has its own accounting and IT unit. The CRM manager is the authorised and accountable person for the proper and efficient functioning of the ECS cell.
  1. **Special Structures – Help Desk**: From CRM viewpoint, Help Desk, an important component at the Branch office, is responsible for the following tasks: information and education to visitors, various conveniences and service issues, recording of contact numbers and grievances, calling selected customers on defined occasions, issue of forms, and recording leads.
  2. **Output (Results)**: By utilising the inputs and CRM process efficiently and effectively, the CRM manager usually gets responses, such as incessant flow of intelligence pertaining to customers' needs, perceptions, and expectations, decreased customer complaints, enhanced application

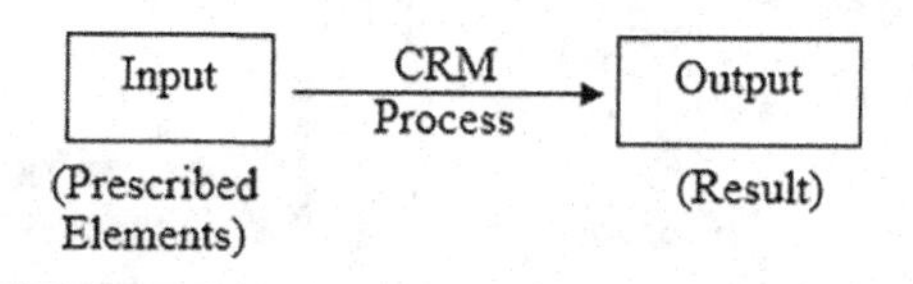

Figure 14.1 CRM process.

of customer portal, accomplishment of quality yardstick in several areas of customer service, growth in renewal premium, constant feedback on process flaws and suggested improvement, alternate channels, IVRS, Info-centre, SMS alerts and such conveniences, retention of customers and repeat sales, and high brand equity, and customer loyalty.

iii. **Redesigned Role of CRE in a Branch**: The CRM concept was introduced in the year 2000 in LIC, and it was envisaged to have a CRE (Customer Relation Executive), preferably an office in the servicing department and Customer Relations Group, comprising 2–3 people in the branches, to have a single point of contact for the customers. The idea behind this was to ensure that the customer is not required to move from seat to seat for any service or grievance. But, it was not effective in implementation, so a new structure has been designed.

iv. **Customer Contacts**: A CRE meets the top customers of the branch and maintains a diary for this purpose. The diary contains the data as outlined in Figure 14.2.

v. **Grievance Redressal**: Resolution of all customer complaints is the fundamental responsibility of the CRE and customer service-related grievances received from all sources, e.g., CMS, IRDAI, Ministry, etc., are taken care of. CRE ensures that no grievances remain unattended beyond a reasonable time. Analysis of complaints and feedback on process improvement, if any, is sent to the CRM manager.

Contact details:

a) Name & address of the customer
b) Phone/e-mail-id
c) Date of Contact
d) Insurance needs
e) Policy profile; No. of policies, premium, product mix
f) Family Insurance profile
g) Service issue, if any and how the same was attended by CRE.
h) Suggestion received
i) Date of Birth of the customer

Follow-up details:

a) New policy taken, if any
b) Policy bond sent within – days
c) Birthday Greeting sent
d) New year greeting sent
e) Festival's greetings sent
f) Calendar/Diary sent
g) Any Maturity/SB due in the year?
h) If so, what special action taken?
i) Any other.

Figure 14.2 Details recorded in the CRE's diary.

vi. **Monitoring the activities of Help Desk**: The Help Desk functions in the branch are enlisted below, and the CRE in the branch is the responsible person for the smooth functioning of this Desk.

- Providing required information to the visitors
- Educating visitors on product/services of LIC
- Recording contact Information
- Recording structural surveys
- Calling special customers on events, such as default, surrender, etc.
- Online registration of the grievances separated at Help Desk
- Generating Quotations, etc., if required by the visiting uniting customer
- Giving acknowledgement for the document received from the customers visiting the branch.

vii. **Monitoring Customer Communication**: A CRE has to perform the following:

- ensure proper and timely dispatch of communications of routine nature, such as notices, default notices, final lapse intimation, ULIP statements
- ensure proper and timely dispatch of some special communications relating to contact programmes designed at the corporate level or customer meets organised at Bo/Do levels
- monitor the overall dispatch of policy bonds within the prescribed time limit.

viii. **Validation of Portal Registration**: After verifying the customer's signature, portal registration is validated. Activities undertaken by the CRE are reported in the enclosed format to CRM manager at the end of every month.

ix. **Renewal Relationship Programme (RRP)**: At the present time, a new programme (RRP) has been introduced in LIC for the existing customers who have not bought any policy for five years after the last policy was bought and who have at least three policies of LIC. The following are the key guidelines under the RRP:

- The customers are contacted over the phone and a suitable timing for personal meeting in cleared.
- It is made clear to the customer that this is not a business call, rather it's a courtesy call and LIC has the intention of strengthening the relationship.
- In case it is found that the customers' loyalties have been shifted, tactful reasons are found out for his/her dissatisfaction with LIC, or liking for other companies. In a competitive market, it is important to understand the relative strengths and weaknesses. Such contacts give an opportunity.

- If cause of dissatisfaction is a grievance, the same is noted and redressed immediately.
- In case the customer is found to be passive for no reason or due to no contact from the LIC side, it is being taken as a business lead and the same may be given to some agent in the branch.

x. **Alternative Premium Payment Channels**: Any facility that helps in retaining customers is part of CRM. LIC provides some alternative premium payment channels, besides visiting the branch, which are in operation and generally accepted by some policyholders, as most of the people of India live in rural areas and they are not aware of this technological development in the LIC. They prefer to pay directly in the branch at the counter or through an agent. Finally, these modes of payment save the precious time of the policyholders. The persons, who avail these facilities, feel happy. The following are some of the important premium payment channels:

- Internet Payment
- ECS
- ATMs of Corporation Bank and AXIS Bank
- Outline collection of premium through AXIS Bank
- Payment through SMS
- Portal

In order to avail the Internet payment facility, a policyholder should have a bank account in India. As at present about seven banks, ICICI, Centurion Bank of Punjab, AXIS, Corporation, HDFC, Federal, and Citibank, are providing online services from the LIC of India side. Nowadays, ECS is one of the most accepted services and is currently available in selected cities where the RBI has widened this service.

To pay premiums, a policyholder needs to have an account in any member bank of the Local Clearing House. The policyholders need not visit a branch for paying the premium or collecting the receipt, once the policyholder has preferred this system of payment. On a particular day, the premium amount will be directly debited to the bank account of the policyholder and the receipt will be issued by the designated branch office as per the instructions given by the policyholder. Apart from that, LIC permits to pay the premium by using the ATMs of Corporations Bank and AXIS Bank for those who have accounts with these basics. Online collection of premium through AXIS Bank, payment through SMS and portal (corporate website, "www.licindia.com") are the other modes of payment that facilitate the payment process.

14.5 Concluding Remarks

BI is a technology-enabled method for evaluating data and disseminating actionable information that enables leaders, managers, and employees to make sound business choices. Organisations collect data from internal and external IT systems and sources, prepare it for analysis, run queries against the data, and create data visualisations, BI dashboards, and reports to make the analytics results available to business users for operational decision-making and strategic planning. The ultimate purpose of BI efforts is to help firms to make more informed business decisions, thereby increasing revenue, increasing operational efficiency, and gaining a competitive edge over competitors. To do this, BI combines analytics, data management, and reporting tools, as well as a variety of data management and analysis approaches.

Life insurance business is one of the most complicated businesses, needs a good relationship, which in turn is based on the BI system. At present, there are 24 life insurance companies in India striving hard for market share. Many life insurers have suffered from customer churning. In particular, LIC of India's market share in terms of renewal premium dipped to 49% in the year 2008 from 100% in the year 2001. As a result, LIC of India examined the causes and focused on BI in general and CRM in particular. However, this was started in 2000 in LIC, but the firm could not implement effectively. Later on, the giant organisation realised and practiced the CRM to hold the existing policyholders. LIC has introduced a new programme, i.e., RRP and redesigned the role of CRE for the effective and smooth functioning of business operations. Some alternative premium payment channels, such as ECS, Internet payment, online payment, ATMs, etc., have been initiated by LIC, which delight the policyholder. It is a fact that LIC has implemented CRM effectively later, as LIC has many awards with regard to customer loyalty and brand equity to its credit. To have a long-lasting competitive advantage over its rivals, LIC of India needs to constantly rely on BI tools in general and CRM in particular.

References

Acheampong, O. & Moyaid, S.A. (2016). An integrated model for determining business intelligence systems adoption and post-adoption benefits in the banking sector. *Journal of Administrative and Business Studies, 2*(2), 84–100.

Alshawi, S., Missi, F., & Irani, Z. (2011). Organisational, technical and data quality factors in CRM adoption—SMEs perspective. *Industrial Marketing Management, 40*(3), 376–383.

Al-Zadjali, M., & Al-Busaidi, K.A. (2018). Empowering CRM through business intelligence applications: A study in the telecommunications sector. *International Journal of Knowledge Management IGI Global, 14*(4), 68–87.

Arunachalam, D., Kumar, N., & Kawalek, J.P. (2018). Understanding big data analytics capabilities in supply chain management: Unravelling the issues, challenges and implications for practice. *Transportation Research Part E: Logistics and Transportation Review, 114*, 416–436.

Athanassopoulos, A.D. (1997). Service quality and operating efficiency synergies for management control in the provision of financial services: Evidence from Greek bank branches. *European Journal of Operational Research, 98*(2), 300–313.

Bakunzibake, P., Grönlund, Å., & Klein, G.O. (2016). E-government implementation in developing countries: Enterprise content management in Rwanda. *15th IFIP Electronic Government (EGOV)/8th Electronic Participation (EPart) Conference*, Univ Minho, IOS Press, Guimaraes, September 5–8, 251–259.

Bagale, G.S., Vandadi, V.R., Singh, D. et al. (2021). Small and medium-sized enterprises' contribution in digital technology. *Annals of Operations Research*. Doi: 10.1007/s10479-021-04235-5.

Bhat, S.A., & Darzi, M.A. (2016). Customer relationship management: An approach to competitive advantage in the banking sector by exploring the mediational role of loyalty. *International Journal of Bank Marketing, 34*(3), 388–410.

Chalmeta, R. (2006). Methodology for customer relationship management. *Journal of Systems and Software, 79*(7), 1015–1024.

Chandrapal, J. D. (2019). Impact of liberalisation on Indian life insurance industry: A truly multivariate approach. *IIMB Management Review, 31*(3), 283–297.

Chen, C. P., & Zhang, C. Y. (2014). Data-intensive applications, challenges, techniques and technologies: A survey on Big Data. *Information Sciences, 275*, 314–347.

Dooley, P., Levy, Y., Hackney, R.A., & Parrish, J.L. (2018). Critical value factors in business intelligence systems implementations. in Deokar, A., Gupta, A., Iyer, L. and Jones, M. (Eds), *Analytics and Data Science. Annals of Information Systems*. Cham: Springer.

Hatta, N. N. M., Miskon, S., & Abdullah, N. S. (2017). Business intelligence system adoption model for SMEs. In *Pacific Asia Conference on Information Systems (PACIS)*, Langkawi. *Association For Information Systems*.

Kannan, A.D., & Vikkraman, P. (2015). Implementation of CRM processes in life insurance sector: A customers' perspective analysis. *Purushartha, 8*(2), 78–84.

Kautish, S. (2008). Online banking: A paradigm shift. *E-Business, ICFAI Publication, Hyderabad, 9*(10), 54–59.

Kautish, S., Singh, D., Polkowski, Z., Mayura, A., & Jeyanthi, M. (2021). *Knowledge Management and Web 3.0: Next Generation Business Models*. Berlin: De Gruyter.

Kautish, S., & Thapliyal, M.P. (2013). Design of new architecture for model management systems using knowledge sharing concept. *International Journal of Computer Applications, 62*(11), 27–30.

Manogaran, G., & Lopez, D. (2017). A survey of big data architectures and machine learning algorithms in healthcare. *International Journal of Biomedical Engineering and Technology, 25*(2–4), 182–211.

Nasab, S.S., Jaryani, F., Selamat, H., & Masrom, M. (2017). Critical success factors for business intelligence system implementation in public sector organization. *International Journal of Information Systems and Change Management, 9*(1), 22–43.

Olszak, C. M., & Ziemba, E. (2007). Approach to building and implementing business intelligence systems. *Interdisciplinary Journal of Information, Knowledge, and Management, 2*(1), 135–148.

Owusu, A., Agbemabiasie, G.C., Abdurrahaman, D.T., & Soladoye, B.A. (2017). Determinants of business intelligence systems adoption in developing countries: An empirical analysis from Ghanaian banks. *The Journal of Internet Banking and Commerce, 24*(2), 1–25.

Popovič, A., Hackney, R., Coelho, P. S., & Jaklič, J. (2012). Towards business intelligence systems success: Effects of maturity and culture on analytical decision making. *Decision Support Systems, 54*(1), 729–739.

Rani, M. (2012). CRM in Insurance services. *International Journal of Innovations in Engineering and Technology, 1*(1), 51–58.

Rygielski, C., Wang, J. C., & Yen, D. C. (2002). Data mining techniques for customer relationship management. *Technology in Society, 24*(4), 483–502.

Sauter, V. L. (2014). *Decision Support Systems for Business Intelligence.* John Wiley & Sons, Hoboken, New Jersey.

Siddiqui, S. A. (2020). Evaluating the efficiency of Indian Life insurance sector. *Indian Journal of Economics and Development, 16*(1), 72–80.

Singh, A., & Gite, P. (2015). Corporate governance disclosure practices: A comparative study of selected public and private life insurance companies in India. *Apeejay - Journal of Management Sciences and Technology, 2*(2).

Singh, D., Singh, A., & Karki, S. (2021). Knowledge management and Web 3.0: Introduction to future and challenges. In *Knowledge Management and Web 3.0.* De Gruyter, Cambridge University Press. Doi:10.1515/9783110722789-001Agents.

Uddin, F. (2020). CRM: A differentiating strategy to retain customer in life insurance sector. *The Management Accountant, ICMAI,* 55(9), 35–37.

Vercellis, C. (2009). *Business Intelligence: Data Mining and Optimization for Decision Making* (pp. 1–18). New York: Wiley.

Villamarín, J.M., & Diaz Pinzon, B. (2017). Key success factors to business intelligence solution implementation. *Journal of Intelligence Studies in Business, 7*(1), 48–69.

Williams, S., & Williams, N. (2006). *The Profit Impact of Business Intelligence.* San Francisco, CA: Morgan Kaufmann.

Yeoh, W., & Koronios, A. (2010). Critical success factors for business intelligence systems. *Journal of Computer Information Systems, 50*(3), 23–32.

Yeoh, W., & Popovič, A. (2016). Extending the understanding of critical success factors for implementing business intelligence systems. *Journal of the Association for Information Science and Technology, 67*(1), 134–147.
Zineldin, M. (2005). Quality and customer relationship management (CRM) as competitive strategy in the Swedish banking industry. *The TQM Magazine.*

Index

Note: **Bold** page numbers refer to tables and *italic* page numbers refer to figures.

Printed in the United States
by Baker & Taylor Publisher Services